A Primer of Higher Space

(The Fourth Dimension)

Claude Bragdon

NEW YORK

A Primer of Higher Space (The Fourth Dimension)
© 2005 Cosimo, Inc.
All rights reserved. No part of this book may be used or reproduced in any manner whatsoever without prior written permission except in the case of brief quotations embodied in critical articles or reviews.
For information, address:

Cosimo
P.O. Box 416
Old Chelsea Station
New York, NY 10113-0416

or visit our website at:
www.cosimobooks.com

A Primer of Higher Space (The Fourth Dimension) originally published by Omen Press in 1972.

Library of Congress Cataloging-in-Publication Data
A catalog record for this book is available from the Library of Congress

Cover design by www.wiselephant.com

ISBN: 1-59605-361-5

TO

PHILIP HENRY WYNNE

Whose pole-true intelligence has from first to last guided the author in this adventure upon uncharted oceans of thought.

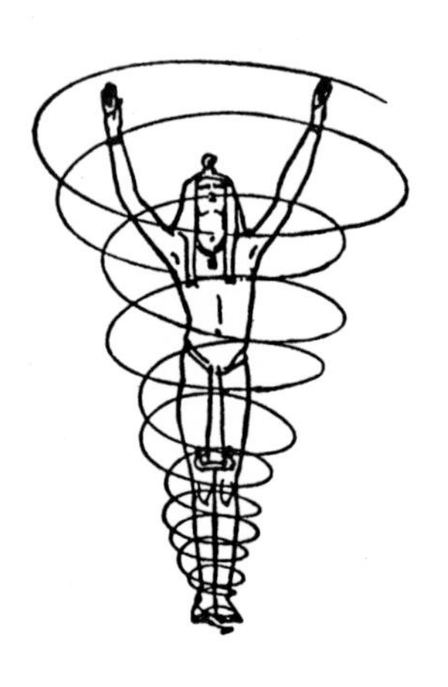

THE FOURTH DIMENSION

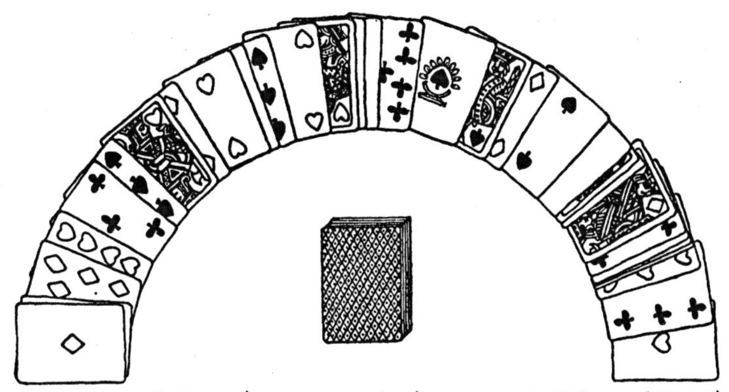

A PACK OF CARDS: A 3-DIMENSIONAL UNITY, BECOME MANIFEST IN THE 2 DIMENSIONS OF A PLANE SPACE

THE FOURTH DIMENSION

THE INDISPENSABLE ANALOGY

ADVENTURE with me down a precipice of thought, sustained only by the rope of an analogy, slender but strong. This rope, anchored in the firm ground of sensuous perception, extends three paces in the direction of the great abyss, then vanishes at the giddy brink. Let us examine this sustaining simile foot by foot and strand by strand.

Familiar both to the mind and eye are the space systems of one, two, and three dimensions; that is, lines, planes, solids. Lines are bounded by points, and themselves bound planes; line-bound planes in turn bound solids. *What, then, do solids bound?* Here is where the analogical rope vanishes from sight. If you answer that a solid cannot be a boundary we part company. No argument of mine can convince you to the contrary. But if you are interested enough to ask, "Well, what *do* solids bound?" logic compels the answer, "*Higher* solids: four-dimensional forms (invisible to sight) related to the solids we know as these are related to their bounding planes, as planes to their bounding lines."

Let us retrace our steps and go over the ground again. A point, moving a given distance in an unchanging direction, traces out a line. This line, moving in a direction at right angles to itself a distance equal to its length, traces out a square. This square, moving in a direction at right angles to its plane a distance equal to the length of one of its sides, traces out a cube. It is easy to picture these processes and the resultant geometrical figures of one, two, and three dimensions, because the line, the square, and the cube have their correlatives in the world of objects; but the imagination fails when the attempt is made to continue this order of form-building. Here again the rope disappears into the void. For the cube to develop in a direction at right angles to its every dimension a new region of space would be required—*a fourth dimension.* Should you declare that there is not and cannot be such a region of space, I wave you farewell, as before. But if you hesitate, I cannot forbear to press my advantage. In such a higher space the cube would trace out a *hyper-cube,* or *tesseract,* a four-dimensional figure related to the cube as the cube is related to the square. This figure, invisible to the eye, is known to the mind. The number of its points, lines, planes and cubic boundaries, and their relation to one another, are as familiar to the mathematician as are the elements of the cube whence it is derived.*

Again return, and for the third time. Arithmetically it is possible to raise a number to any given power; that is, to multiply it by itself any number of times. There is a known spatial correlative of the second power of a number, the square; and of its third power, the cube; but we have no direct or sensory knowledge of that analogous form, the tesseract, which would correspond to the fourth power of a number, nor of the four-dimensional space in which alone its development would be possible. With the geometry of such a space mathematicians long have been familiar, but is there such a space—is there any body for this mathematical soul?

THE REASONABLENESS OF THE HIGHER SPACE HYPOTHESIS

Before dismissing such an idea as absurd, let it be remembered that the mathematician is today the scout of science. Of this fact the discovery of Neptune is a classic example. A French mathematician computed and announced the place of a hypothetical body exterior to Uranus. A German astronomer pointed his telescope towards the designated quarter of the heavens and found an object with a planetary disc not plotted on the map of stars. It was the sought for world. May not the preoccupation of mathematicians with problems involving a hypothetical space of four dimensions anticipate the discov-

*The hyper-cube has 16 corners, 32 edges, 24 square faces and 8 bounding cubes.

ery and conquest, not of a new world, but of a new space?

The existence of a space of four dimensions can never be disproven by showing that it is absurd or inconsistent, for such is not the case. Not only has the higher space hypothesis such validity as the great law of analogy can give, but it solves many of the problems and reconciles many of the contradictions which confront the modern man of science. So true is this that Helmholz continually kept the possibility of physical higher space before him in his dynamical reasonings which are among the classics of physical science. Kelvin also felt the pressure of the mathematical reality of higher space so strongly that he declared himself ready to accept it as an explanation of physical phenomena when these could be more consistently explained by such a concept.

The idea that space may have more than three dimensions may seem highly revolutionary, but it is no more so than the idea that the world is spherical instead of flat, or that the earth revolves around the sun instead of the sun around the earth. These familiar and established truths contradict the evidence of the senses, and therefore they were received with incredulity until the proof of them became overwhelming. Should the evidence in favor of a many-dimensional space become overwhelming, we should be forced to accept the idea and come to think in terms of it. In other words, if we came to observe in space contradictory facts, and if these facts forced us to ascribe to a body two attributes or qualities which we formerly thought could not exist together, our reason would seek to reconcile these contradictions. If the idea of a fourth dimension reconciled them, we should develop a sense of higher space.

THE FOURTH DIMENSION DEFINED

The expression, *the fourth dimension,* offers a shock to the mind accustomed to practical handling of matter, because all our experiences of measurement or dimensionality are ultimately founded upon matter possessing but three dimensions, so that we have great difficulty in accepting the reality of a direction not contained in our space or our matter but definitely at right angles to every line that can be drawn within the matter and space which contain all our ordinary experiences. Our idea of space is partial, and like many another of our ideas needs modification to accommodate it to fuller knowledge. What we think of as space is more probably only some part of *space made perceptible.* It may be that our space bears a relation to space in its totality analogous to that which the images cast by a magic lantern bear to the wall on which these images are made to appear—a wall with solidity, *thickness,* extension in other and more directions than those embraced within the wavering circle of light which would correspond to our sense of the cosmos. In other words, perhaps that which we think of as space is only so much of it as our limited sensuous mechanism is able to apprehend.

For knowledge arises from consciousness, and consciousness everywhere and always is conditioned by the vehicle of physical perception. The particular "space" or dimensional order apprehended by consciousness must bear an exact relation to the amplitude of motion *in* space of which the vehicle is capable. As this amplitude of motion varies widely in different departments of nature, there naturally arises the idea of spaces of different dimensionalities, each added dimension corresponding to a power of motion in a new direction. The grub, working its way upward out of the earth in which it is buried, may be said to inhabit a linear, or one-dimensional space; the caterpillar's space, the surface of a leaf, is two-dimensional; while the winged butterfly attains the freedom of all three dimensions. Understood in this way—as new powers of movement in new media—the expression *the fourth dimension of space* is sufficiently descriptive of an unfamiliar power of movement in an unknown medium, but related to the movements and the media known to us by an orderly sequence of evolution.

THE DESIGNATION OF THINGS IN TERMS OF THEIR DIMENSIONALITY

One other apparent contradiction in the use of terms, involving the possibility of misconception, should also be mentioned. In any discussion of the higher space hypothesis such expressions as a "one-dimensional" body, or a "two-dimensional" body, are apt to occur. Such terms, though convenient, are not accurate, because there can be no such thing as a one-dimensional or a two-dimensional body in our three-dimensional space. Its three dimensions must exist always and everywhere, and everything must have some extension, however slight, in every one of its three dimensions. This is clear, even axiomatic, and the case must be the same if space has more than three dimensions. *Whatever the number of dimensions in space, everything must have that number of dimensions, too.* If there are more than three dimensions in space and we perceive only three in the objects around us, then we are forced to the conclusion that we perceive things only partially. All the dimensions beyond the third are *for us* non-existent, but our lack of perceptive power does not in any way affect the objects themselves.

Notwithstanding this self-evident fact (that the number of dimensions in an object must be co-equal with the number of dimensions in space itself), there is a sense in which such expressions as a "one-dimensional" or a "two-dimensional" body have quite sufficient validity. They may be used to designate, first, a cross-section, limit, or boundary—the edge of a razor would be one-dimensional in this sense, and the surface of a table two-dimensional—and second, to designate something in which the given number of dimensions is patent, and any dimensions beyond these latent; or in which the constituent particles have, or appear to have, free movement or power of transmission in the designated number of dimensions, and restricted movement in all directions above and beyond that number. Understood in this way, a nerve, a hair, the stem of a leaf—the trunk of a tree, even—might properly be called one-dimensional; a leaf, a handkerchief, a piece of paper, two-dimensional; and any solid of our space three-dimensional. For there is a perfectly appreciable difference, based upon extension in space, between these various classes of objects, notwithstanding that all of them have, as we know, extension in all three dimensions—and according to the higher space hypothesis, in more than that number.

A BROADER CONCEPT OF SPACE

Having now in a manner paved the way, let us put aside all our preconceived ideas as to the limitations of space, and form a new concept which will embrace the higher dimensions as easily as the lower. It is necessary to do this, not because our space conception is false, but only because it is partial. Let us think, not of abstract space, but of material *spaces,* differentiated from one another by their dimensionality, and designated in terms of it (as a one-space, a two-space, a three-space, a four-space, and so on)—the greater the number of dimensions, the "higher" the space. Let us think of each space as generated from the one next below it, and as having the dimensionalities of all spaces lower than itself patent, and those higher than itself latent. Also, conceive of each space as the *cross-section* of the next higher space—as limiting two contiguous portions of higher space. For example, one segment of a line (a one-space) is divided from another by a point, and the line itself is generated by the motion of a point; one portion of a plane (a two-space) is separated from another by a line (a one-space), and the plane itself is generated by the movement of the line in a direction at right angles to its length. Again, two portions of a solid (a three-space) are limited with regard to one another by a plane (a two-space); this plane, moving in a direction at right angles to the surface, generates a solid (a three-space). Also, by analogy, two portions of a higher solid (a four-space) are limited with regard to one another by a solid (a three-space), and this solid, moving in a direction at right angles to its every dimension, generates a higher solid.

From this it is possible to formulate a definition applicable to a space of any dimensionality: *A space is that which separates two portions of the next higher space from each other.* Also, *Any space can generate its next higher space by moving in a new direction, that is, a direction not contained within itself.*

A HYPOTHETICAL TWO-DIMENSIONAL UNIVERSE

With these axioms as a guide, let us by a simple expedient try now to gain some elementary notion of four-dimensional space, undisturbed by any apparent absurdity in the idea. Consider our three-dimensional space as the higher space of a two-dimensional world of similar form and constitution to our own.

Now by the terms of our definition, a two-dimensional universe would be a plane separating two contiguous portions of universal three-dimensional space from each other. This plane, to have concrete existence, to be anything more than a geometrical abstraction, would need to have some thickness, some extension, however slight, in the third dimension. A terrestrial world in such a two-dimensional universe would not be a sphere, as in our three-space, but a circular disc, the cross-section of a sphere. The matter of this disc-world would have the power of free movement in the two extended directions of the plane (its space), but no power of movement in the infinitesimal, or third dimension. Assume that this disc-world is held together, like our own, by an attractive force analogous to gravitation, which not only determines and preserves its circular form, but holds its inhabitants, two-space "men," to the rim which forms its "surface." Such a two-space man would bear the relation to a human being that a cross-sectional slice of a solid bears to the solid itself. What account would this flat-man, on the rim of his disc-world, traversing his plane space, give of the universe in which he finds himself?

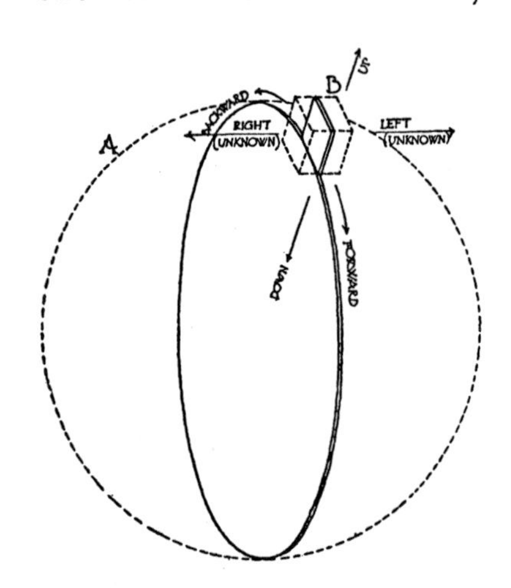

The direction of the attractive force holding him to the rim of his disc (the surface of his earth) would give him the sense of up and down, the airy region above his head and the earthy region beneath his feet, thus determining for him one dimension of his plane space. Also, since he can move forward and backward along that rim, he would be conscious of a direction parallel to the circumference—the second dimension; but being denied any further power of movement he would have no knowledge of right or left—the third dimension—for this direction would extend into his "higher" or three-dimensional space. Suppose him to be inspired by some intimation of this higher space (as we are beginning to be inspired by intimations of a space higher still), in what form would the apprehension of it enter his limited consciousness, and how would it affect his concepts?

He would come to conceive that the conditions of his existence demanded the presence of an extended wall of matter everywhere in contact with the objects of his world. He would infer also that these objects must have an imperceptible thickness in the higher dimension—that they might indeed conceivably (though invisibly to him) extend away into it. What seemed to him to be the solids of his world might be, in fact, the two-dimensional boundaries or cross-sections of three-dimensional solids in, or passing through his space, in the same manner that in his world a one-dimensional line would be recognized as the boundary or cross-section of a two-dimensional square. Reasoning along these lines, he might conclude that he himself was but a two-dimensional cross-section of some three-dimensional body, the corporeal part of a higher being inhabiting three-dimensional space; and the sublime faith might grow up in him that his unified consciousness had its home in planes of existence the nature of which, by his physical limitation to two dimensions, he could not be aware.

DEDUCTIONS FROM FOREGOING PREMISES

If this train of reasoning is logical for a two-space man, exactly analogous suppositions may be formed by us with regard to four-dimensional space.

We have a right to infer that our space is, as it were, an interval, a gap, in higher space. We may believe that there is a direction extending at right angles to every direction that we know; that our world and everything in it is sustained and environed by this boundary. We must, however, give up any attempt to picture to our imagination this four-dimensional space. We can no more do so than the flat-man can imagine the three-dimensional space contiguous to his own. We cannot point to it any more than the two-space man can point to his right or to his left, because it involves a power of movement which we do not possess. We may suppose that in some way unknown to us all the objects of our world have an infinitesimal "thickness" in the higher dimension; that they are in reality three-dimensional projections or cross-sections of higher solids traversing our space. We may conclude, indeed, that our physical bodies are such projections of higher bodies—still our own—and that our essential selves have their home in planes of being, the nature of which, by the limitation of our consciousness to three dimensions, we cannot be aware.

It may be urged that the cross-sectional space, world, man, first described, has nowhere any existence, that the whole concept is a figment of the imagination, and that conclusions based upon premises of this order are entitled to no consideration. It is true that we know of no two-dimensional space, world, or beings such as have been described; the whole analogy is an artifice, a false-work, constructed for the purpose of acquainting the reader, easily and rapidly, with certain fundamental higher space concepts. But these conclusions, though deduced from premises avowedly artificial, are not for that reason necessarily false. They are not solely sustained by means of this flimsy scaffolding, but rest upon other and firmer foundations, as will be shown.

A PHYSICAL TWO-DIMENSIONAL SYSTEM

And yet this particular scaffolding is not quite so rickety as it appears. A study of nature from the standpoint of the higher space theory—that is, the consideration of it as an aggregation of material spaces of different dimensionalities—brings to light many systems resembling in their essential elements the hypothetical one above sketched. Recollect first that space, as the word is here used, is the space of the physicist rather than that of the geometer. Space should be conceived of as concrete and relative, not as abstract and absolute. With this idea in mind, and granting that all life manifesting as motion is conscious life, as much aware of its environment as the limitation of its motion permits, let us see if we cannot find some combination of related spaces of different dimensionalities—actual, concrete spaces—which shall form a basis for deductions of the same order as those already made from imaginary premises.

Since by the terms of our definition, "A space is that which limits two contiguous portions of the next higher space," the surface of a pond or pool, limiting as it does the water beneath and the air above, would constitute a two-dimensional space limiting two portions of three-dimensional space from each other. A lily-pad floating horizontally on this surface might be considered a world of this two-space, and its inhabitant some microscopic unit of consciousness free to move in the two extended dimensions of the leaf, but prevented by the *thinness* of its plane world from moving up and down, and therefore necessarily ignorant of everything above and below. Except for a change of axis, this corresponds in its essential features to the cross-sectional system first sketched, and nature abounds in such systems and sequences. The plane space, plane world, and plane-bound consciousness so far from being arbitrary and absurd, are literally true in fact.

THE MOVABLE THRESHOLD OF CONSCIOUSNESS

Du Prel says, in his *Philosophy of Mysticism,* "From the standpoint of every animal organism we can divide external nature into two parts, which are the more unequal as the organic grade is lower, the one includes that part with which the sense apparatus establishes relations; the other is transcendental for the organism in question; that is, the organism lives in no relation to it. In the biological process the boundary line between these two world-halves has been pushed continually forward in the same direction. The number of senses has increased, and their functional ability has risen. . . The biological rise and the rise of consciousness thus signify a constant removal of the boundary between representation and reality at the cost of the transcendental part of the world, and in favor of the perceived part."

Now if this shifting psycho-physical threshold is simply the dividing line between lower and higher spaces, then the whole evolutionary process consists in the conquest, dimension by dimension, of successive space-worlds. This certainly holds true as far as our observation extends. To the grub, working its way up to the surface of the earth, that surface is transcendental; to the caterpillar, the earth is real, and the free air transcendental; while to the butterfly, master of this added dimension, the threshold has again receded. Indeed there are indications that the butterfly is in possession of a space-sense which is still a mystery to us. Fabre himself cannot explain how the great peacock moth finds its mate in the dark and at a distance sometimes of miles.

Arguing by analogy, everything which is to us transcendental exists nevertheless in some space. It is therefore possible that by an intention of consciousness we may be able first to apprehend, then to perceive as real, that which is now considered transcendental.

THE CONQUEST OF SPACE BY CONSCIOUSNESS

Right at this point we cross the track of Plato. In one of the Platonic dialogues Socrates makes an experiment on a slave who is standing by. He causes space perceptions to awaken in the mind of Meno's slave by directing his close attention to some simple facts in geometry. Plato's comment was that behind the phenomena of mind that Meno's slave boy exhibited, there was a vast, an infinite perspective. It is possible that by dwelling on elementary higher space concepts, we may repeat Socrates' experiment on new grounds; that is, by an intention of consciousness upon the fourth dimension we may push back the psycho-physical boundary and capture for sense the now transcendental fourth dimension.

Because this pushing back of the psycho-physical boundary is incessant and universal, we have only to observe its lower and earlier manifestations in order to understand its immediate and ultimate. Such observation leads to the conclusion that our sense of time may be only an imperfect sense of space. This is clear if we consider the manner in which the facts of space must present themselves to a consciousness with a less developed sense of space than ours—and by a less developed sense of space is meant a more limited power of representation in form. Many things which we as human beings apprehend without difficulty *simultaneously,* after the manner of space, such a consciousness could only apprehend *successively* after the manner of time. A worm, for example, requires time in order to examine an angle or a hole—things in which with us the time element does not enter at all. Thus, that which is *time* to one grade of consciousness, is *space* for the next ascending grade. If this is true, then though the fourth dimension cannot manifest itself to our three-dimensional powers of perception as space, it can, and perhaps does, manifest itself as time—that is, by means of changes of state in the objects of our world, involving a temporal element.

THE FOURTH DIMENSION AS TIME

This is important. Let us see if we cannot realize it more completely and concretely. Think of the fourth dimension, not as a new region of space —a direction, as has been said, towards which we can never point—but as a principle of growth, of change, a measure of relations which cannot be expressed in terms of length, breadth and thickness. Now go back to the consciousness limited to the two dimensions of a plane. The objects of our world (three-dimensional objects) passing through his plane world would manifest their third dimension as a principle of growth, of change, and as a measure of relations which could not be expressed in terms of two dimensions. The changing cross-sections which they traced in the constituent matter of his plane in passing through would alone be in evidence. These, appearing, waxing, waning, would seem to him to be the matter of his world in a dynamic condition, organized into forms mysteriously endowed with an inherent power of change, of expansion and contraction. Imagine, for example, a cone passed, apex downward, through a plane. It would appear there first as a point, expanding into a circle, and this still growing circle would suddenly disappear. All of these modifications of form would be caused by the gradual involvement of the third dimension of the cone into the two dimensions of the plane—they would be a temporal expression of the cone's extension in that third dimension. This is a commonplace of science. All higher dynamical reasonings use motion as a translator of time into space or space into time.

If, then, as thus appears, the third dimension could manifest itself to a consciousness limited to two dimensions as a sequence of changes in two-dimensional objects which required time for their unfoldment; then, by analogy, a fourth dimension, which would be spatial extension in some new direction at right angles to the three known to us, would manifest itself to our perception equally as a time change. What changes, involving a temporal element might be regarded as significant of higher dimensionality? What but the universal flux of things—life, growth, organic being, the transition from simplicity to complexity, the shrinkage or expansion of solids? These would be, by this view, the evidence of a fourth dimension; they would be measurable only by means of it, since a temporal element is involved in every such change.

One manner of conceiving the fourth dimension, therefore, is as *space changing in time.* We are to think of the physical universe accessible to our observation as possessing at least four co-ordinate and interchangeable dimensions, of which three are included under the name of space, and the fourth is called time. If all movement in space were suddenly to cease, the fourth dimension would be eliminated from it. Fantastic as this idea may appear, it is exactly that which has produced interesting results in dealing with the problem presented by the ether of space. Mathematical physicists have found that apparent experimental contradictions disappear and the mathematical framework of physics is greatly simplified if, instead of referring phenomena to a set of three space axes and one time axis of reference, they are referred to a set of four interchangeable axes involving four homogeneous co-ordinates, three of space and one of time. Time, in other words, is employed as though it were a dimension of space—the fourth dimension.

THE HISTORY OF HIGHER SPACE THOUGHT

While the higher space hypothesis is a flower of the modern garden, it has its roots in the rich soil of the past. No great violence of invention is required to discover (as has been done) the idea of higher space in the Ancient Wisdom of India, and in the philosophical systems of Parmenides and Pythagoras. As C. Howard Hinton says, "Either one of two things must be true—that four-dimensional conceptions give a wonderful power of representing the thought of the East, or that the thinkers of the East must have been looking at and regarding four-dimensional existence." Plato, in a famous passage in the *Republic,* shows that he held the precious secret in the hollow of his hand. Reference is made to the memorable allegory of the chained captives, reduced to be the denizens of a shadow world. All movements observed by them were but movements on a surface, and all shapes but the shapes of outlines with no substantiality. Plato uses this illustration to portray the relation between true being and the illusions of the sense world. It is a significant fact that the term "quarta dimensio" was used first by Henry More, the Platonist, about the year 1671.

Kant not only recognized the possibility of the existence in space of more than three dimensions, but he inferred their very probable real existence. "If it is possible," he says, "that there are developments of other dimensions of space, it is also very *probable* that God has somewhere produced them. For His works have all the grandeur and variety that can possibly be comprised." Swedenborg's involved descriptions of "heavenly" forms, motions and mechanics, become somewhat more intelligible when interpreted in terms of higher space. It would appear that in common with many other seers he knew only as much of four-dimensional existence as fish know of water—they are unconscious of it because it is the medium in which they live and move. Certain modern researchers in these super-physical realms are aware of the unique character of their environment. One of them, Mr. C. W. Lead-beater says, "I can at any rate bear witness that the tesseract, or four-dimensional cube, is a reality, for it it is quite a familiar figure on the astral plane."

Modern science, to which testimony of this order has no validity because it is not susceptible of corroboration by the usual methods, has approached the subject of higher space along totally different lines. The notion of geometries of n dimensions—the geometry, that is, of higher space—began to suggest itself to mathematicians in the early half of the nineteenth century, and has assumed an increasing importance ever since. Attempts to utilize the higher space hypothesis in the explanation of chemical and physical phenomena has served to bring the subject prominently before workers in these branches of science. And finally, because the hypothesis would account for many so-called psychic phenomena, it has been seized upon by the psychic researcher, hard pressed for some quasi-scientific explanation of things the reality of which he cannot doubt.

Research leads always into the profound. The light of things known serves but to reveal a greater abysm of mystery beyond the threshold of consciousness. The higher space hypothesis makes man in his present estate appear but as an earthworm in power and knowledge, nevertheless it holds out the promise of eternal progress.

PLATES

THE GENERATION OF CORRESPONDING FIGURES IN ONE-, TWO-, THREE-, AND FOUR-SPACE.

FIG. 1.

THE LINE: A 1-SPACE FIGURE GENERATED BY THE MOVEMENT OF A POINT; CONTAINING AN INFINITE NUMBER OF POINTS, AND 2 FORM ITS BOUNDARIES

FIG. 2.

THE SQUARE: A 2-SPACE FIGURE GENERATED BY THE MOVEMENT OF A LINE IN A DIRECTION PERPENDICULAR TO ITSELF TO A DISTANCE EQUAL TO ITS OWN LENGTH IT CONTAINS AN INFINITE NUMBER OF LINES. AND IS BOUNDED BY 4 LINES AND 4 POINTS.

FIG. 3

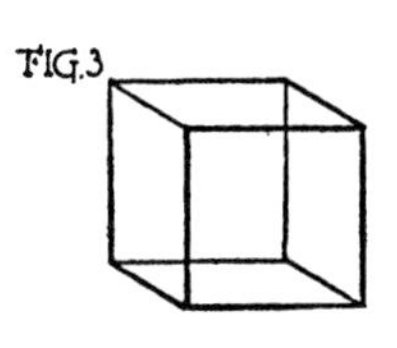

THE CUBE: A 3-SPACE FIGURE OR "SOLID." GENERATED BY THE MOVEMENT OF A SQUARE, IN A DIRECTION PERPENDICULAR TO ITS OWN PLANE, TO A DISTANCE EQUAL TO THE LENGTH OF THE SQUARE THE CUBE CONTAINS AN INFINITE NUMBER OF PLANES (SQUARES) AND IS BOUNDED BY 6 SURFACES, 12 LINES AND 8 POINTS

FIG. 4

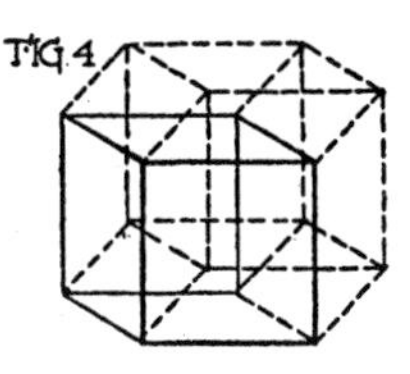

FIG. 5.

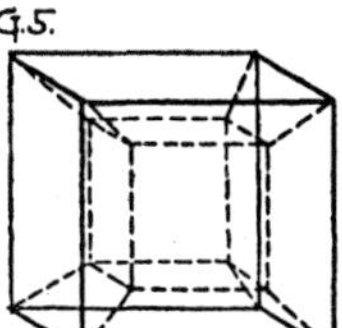

THE TESSERACT, OR TETRA-HYPERCUBE: A 4-SPACE FIGURE GENERATED BY THE MOVEMENT OF A CUBE IN THE DIRECTION (TO US UNIMAGINABLE) OF THE 4TH DIMENSION. THIS MOVEMENT IS EXTENDED TO A DISTANCE EQUAL TO ONE EDGE OF THE CUBE AND ITS DIRECTION IS PERPENDICULAR TO ALL OUR 3 DIMENSIONS AS EACH OF THESE 3 IS PERPENDICULAR TO THE OTHERS. THE TESSERACT CONTAINS AN INFINITE NUMBER OF FINITE 3-SPACES (CUBES) AND IS BOUNDED BY 8 CUBES, 24 SQUARES, 32 LINES AND 16 POINTS.

NOTE: FIGURE 4 IS A SYMBOLIC REPRESENTATION ONLY—A SORT OF DIAGRAM—SUGGESTING SOME RELATIONS WE CAN PREDICATE OF THE TESSERACT. FIGURE 5 IS A REPRESENTATION DRAWN ON A DIFFERENT PRINCIPLE IN ORDER TO BRING OUT A DIFFERENT SET OF RELATIONS.

CORRELATIONS, IN FORM AND SPACE, OF SOME PROPERTIES OF ABSTRACT NUMBER

THE 3 LINEAR UNITS

FIG. 1.

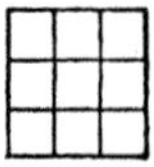

SIDE OF SQUARE 3 <u>LINEAR</u> UNITS. AREA OF SQUARE 9 <u>SQUARE</u> UNITS

A NUMBER MULTIPLIED BY ITSELF GIVES THE SECOND POWER OF THAT NUMBER, COMMONLY CALLED ITS "SQUARE" BY REASON OF ITS CLOSE RELATION TO THE GEOMETRICAL SQUARE WHOSE SIDE CONTAINS THE GIVEN NUMBER OF UNITS OF LENGTH.

THE SECOND POWER OF 3 IS 3 TIMES 3, OR 9. FIG 1 REPRESENTS A GEOMETRICAL SQUARE WHOSE SIDE IS 3 UNITS IN LENGTH, SAY 3 INCHES. THE AREA OF THE SQUARE WILL OBVIOUSLY BE 9 SQUARE INCHES.

FIG. 2

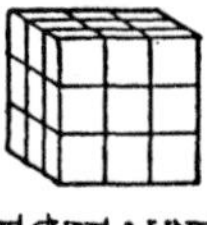

EDGE OF CUBE 3 <u>LINEAR</u> UNITS. FACE OF CUBE 9 <u>SQUARE</u> UNITS. VOLUME OF CUBE 27 <u>CUBIC</u> UNITS

LET US NOW BUILD UP FROM THE SQUARE TO A HEIGHT OF 3 INCHES THE CUBE REPRESENTED IN FIG 2. THE SOLID WILL OBVIOUSLY CONTAIN $3 \times 3 \times 3 = 27$ <u>CUBIC</u> INCHES.

BY ANALOGY WITH THE GEOMETRICAL FIGURE THE NUMBER 27, THE 3RD POWER OF 3, IS CALLED IN ARITHMETIC THE "CUBE" OF 3.

NOW THE 4TH, 5TH AND HIGHER POWERS OF A NUMBER ARE COMMONPLACES OF ARITHMETIC. WHAT DO SUCH HIGHER POWERS MEAN IN GEOMETRY?

WE CANNOT MAKE COMPLETE PHYSICAL REPRESENTATION OF 4-DIMENSIONAL SOLIDS IN OUR 3-SPACE, JUST AS WE CANNOT CONSTRUCT A CUBE IN A PLANE SURFACE BUT WE <u>CAN</u> MAKE DIAGRAMS OF HYPERSOLIDS, AND THE PROPERTIES OF MANY SUCH FIGURES IN HYPERSPACE ARE WELL KNOWN, HAVING BEEN DEMONSTRATED LIKE PROPOSITIONS IN EUCLID.

HYPERSPACE IS THUS <u>MATHEMATICALLY</u> <u>REAL</u>, AND THE MASTER MINDS OF SCIENCE CONSIDER IT TO BE <u>PHYSICALLY</u> <u>POSSIBLE</u> (LORD KELVIN AND OTHERS).

A 2-SPACE UNIT

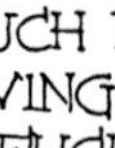

A 3-SPACE UNIT

A 4-SPACE UNIT

THE DEVELOPMENT OF A UNIT OF 2, 3, AND 4 SPACE INTO THE NEXT LOWER SPACE AND THEIR EXPRESSION IN AND BY MEANS OF UNITS OF THOSE LOWER SPACES

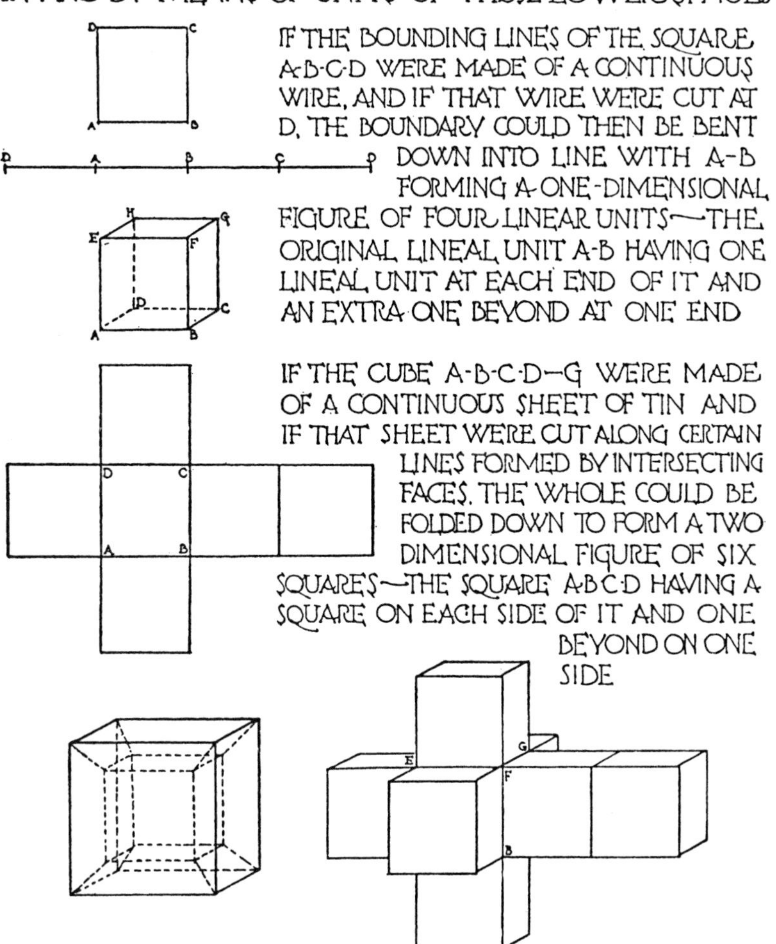

IF THE BOUNDING LINES OF THE SQUARE A-B-C-D WERE MADE OF A CONTINUOUS WIRE, AND IF THAT WIRE WERE CUT AT D, THE BOUNDARY COULD THEN BE BENT DOWN INTO LINE WITH A-B FORMING A ONE-DIMENSIONAL FIGURE OF FOUR LINEAR UNITS—THE ORIGINAL LINEAL UNIT A-B HAVING ONE LINEAL UNIT AT EACH END OF IT AND AN EXTRA ONE BEYOND AT ONE END

IF THE CUBE A-B-C-D—G WERE MADE OF A CONTINUOUS SHEET OF TIN AND IF THAT SHEET WERE CUT ALONG CERTAIN LINES FORMED BY INTERSECTING FACES, THE WHOLE COULD BE FOLDED DOWN TO FORM A TWO DIMENSIONAL FIGURE OF SIX SQUARES—THE SQUARE A-B-C-D HAVING A SQUARE ON EACH SIDE OF IT AND ONE BEYOND ON ONE SIDE

SIMILARLY IF THE TESSERACT (REPRESENTED BY THE DIAGRAM) WERE MADE OF SOLID WOOD AS TO ITS BOUNDING CUBES AND IF THIS WOOD WERE CUT THROUGH THE APPROPRIATE PLANES, THE CUBES COULD, BY ANALOGY, BE FOLDED DOWN TO FORM A THREE DIMENSIONAL FIGURE OF EIGHT CUBES

PLATE 3

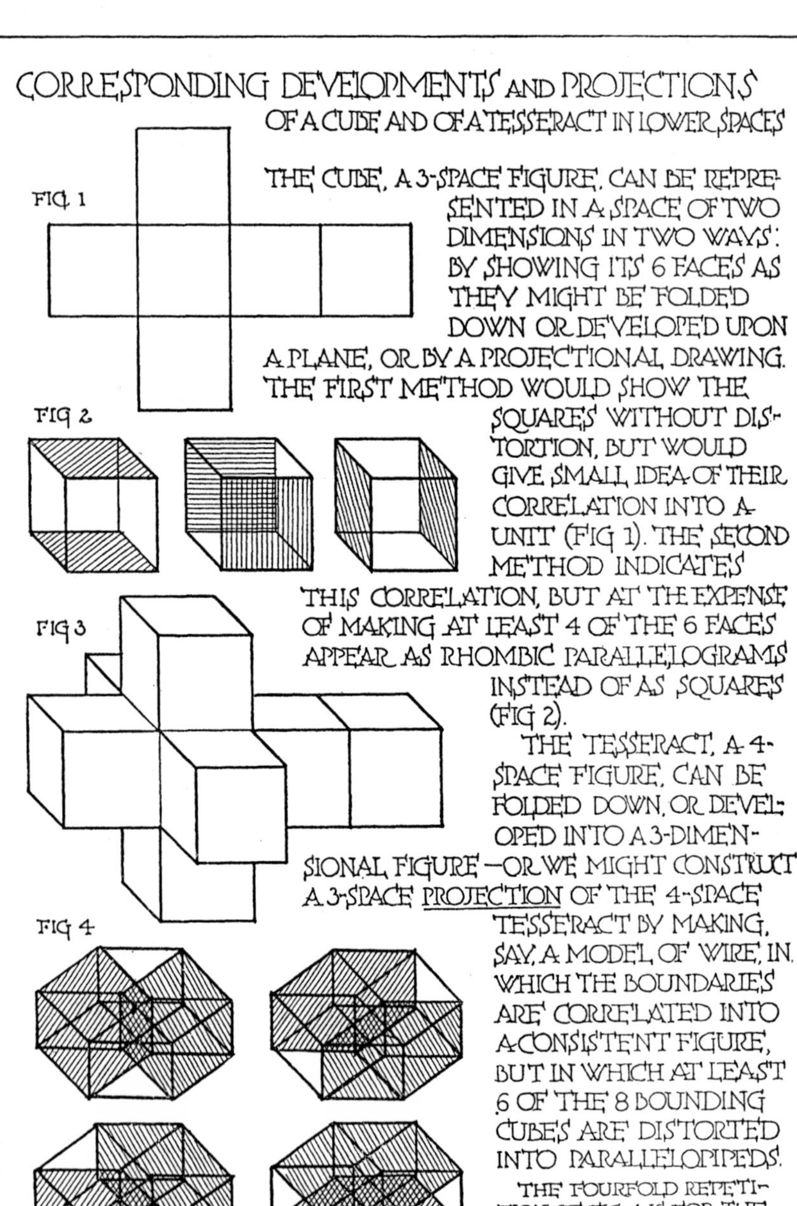

CORRESPONDING DEVELOPMENTS AND PROJECTIONS OF A CUBE AND OF A TESSERACT IN LOWER SPACES

THE CUBE, A 3-SPACE FIGURE, CAN BE REPRESENTED IN A SPACE OF TWO DIMENSIONS IN TWO WAYS: BY SHOWING ITS 6 FACES AS THEY MIGHT BE FOLDED DOWN OR DEVELOPED UPON A PLANE, OR BY A PROJECTIONAL DRAWING. THE FIRST METHOD WOULD SHOW THE SQUARES WITHOUT DISTORTION, BUT WOULD GIVE SMALL IDEA OF THEIR CORRELATION INTO A UNIT (FIG 1). THE SECOND METHOD INDICATES THIS CORRELATION, BUT AT THE EXPENSE OF MAKING AT LEAST 4 OF THE 6 FACES APPEAR AS RHOMBIC PARALLELOGRAMS INSTEAD OF AS SQUARES (FIG 2).

THE TESSERACT, A 4-SPACE FIGURE, CAN BE FOLDED DOWN, OR DEVELOPED INTO A 3-DIMENSIONAL FIGURE—OR WE MIGHT CONSTRUCT A 3-SPACE PROJECTION OF THE 4-SPACE TESSERACT BY MAKING, SAY, A MODEL OF WIRE, IN WHICH THE BOUNDARIES ARE CORRELATED INTO A CONSISTENT FIGURE, BUT IN WHICH AT LEAST 6 OF THE 8 BOUNDING CUBES ARE DISTORTED INTO PARALLELOPIPEDS.

THE FOURFOLD REPETITION OF FIG 4 IS FOR THE PURPOSE OF EXHIBITING THE 8 BOUNDING CUBES OF THE TESSERACT

PLATE 4

THE REPRESENTATION AND ANALYSIS OF THE TESSERACT, OR FOUR-DIMENSIONAL CUBE BY A METHOD ANALOGOUS TO THAT EMPLOYED IN MAKING A PARALLEL PERSPECTIVE

FIG 1.

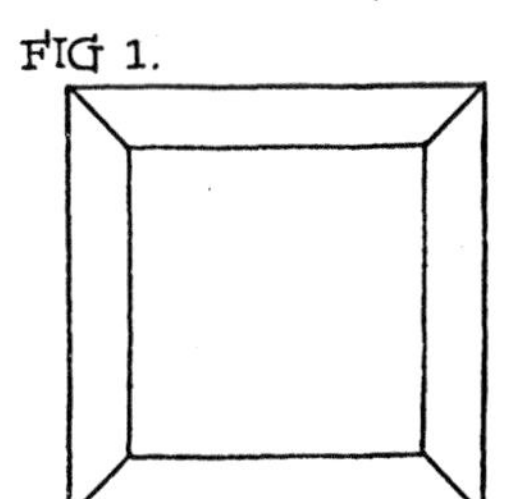

A GLASS CUBE, HELD DIRECTLY IN FRONT OF THE EYE, WILL APPEAR AS SHOWN IN THE ACCOMPANYING DRAWING. THIS—BEING A PLANE FIGURE OF TWO DIMENSIONS—MIGHT HAVE BEEN PRODUCED BY DRAWING ONE SQUARE INSIDE OF ANOTHER AND THEN CONNECTING THE CORRESPONDING CORNERS. THIS COULD BE DONE WITHOUT ANY THOUGHT OF THREE DIMENSIONS. YET ON THIS PLANE FIGURE MANY OF THE PROPERTIES OF THE CUBE CAN BE STUDIED. BY COUNTING THE FOUR-SIDED FIGURES, WHICH WE FIND TO BE SIX, WE LEARN THE NUMBER OF FACES OF THE CUBE. BY COUNTING THE NUMBER OF CORNER POINTS, WHICH ARE EIGHT, WE LEARN THE NUMBER OF THE CORNERS OF THE CUBE. BY COUNTING THE LINES, WHICH ARE TWELVE, WE LEARN THE NUMBER OF EDGES OF THE CUBE.

FIG 2.

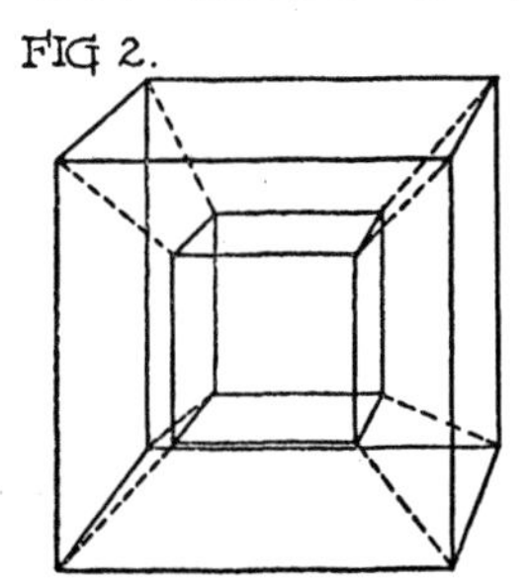

IN THE SAME WAY THAT FIGURE 1 REPRESENTS THE CUBE, FIGURE 2 REPRESENTS THE THREE-DIMENSIONAL FORM CORRESPONDING TO THE TESSERACT. JUST AS WE DREW A SMALLER SQUARE INSIDE OF A LARGER ONE, SO WE REPRESENT A SMALLER CUBE INSIDE OF A LARGER CUBE, AND JUST AS WE DREW LINES JOINING THE CORRESPONDING CORNERS OF THE SQUARES, SO WE FORM PLANES JOINING THE CORRESPONDING EDGES OF THE CUBES. TO FIND THE NUMBER OF CUBIC BOUNDARIES OF THE TESSERACT, WE COUNT THE LARGE OUTER CUBE, THE SMALL INNER CUBE, AND THE SIX SURROUNDING SOLIDS—EACH A DISTORTED CUBE—EIGHT IN ALL. A FURTHER STUDY OF THE FIGURE DISCOVERS 24 PLANE SQUARE FACES, 32 EDGES, 16 CORNER POINTS.

THE BOUNDARY BETWEEN TWO ADJACENT PORTIONS OF ANY SPACE IS, IN GENERAL, A SPACE OF DIMENSIONS FEWER BY ONE —

FOR THE DISCUSSION OF PHYSICAL REALITIES WE ARE TO CONCEIVE EACH KIND OF SPACE AS POSSESSING AN INFINITESSIMALLY SMALL EXTENSION IN THE NEXT HIGHER DIMENSION

FIG 1

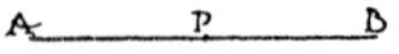

A POINT, OR 0-SPACE DIVIDES A LINE, OR 1-SPACE INTO TWO PARTS (FIG 1) FOR OUR PHYSICAL REASONINGS WE MAY TAKE FOR THE POINT A CIRCLE OF 1/200,000,000 OF AN INCH DIAMETER — ABOUT THAT OF A MOLECULE.

FIG 2.

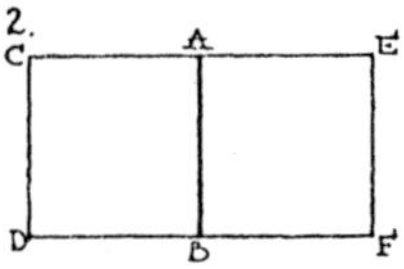

THE LINE A—B (FIG 2), A 1-SPACE, FORMS THE BOUNDARY BETWEEN TWO ADJACENT PARTS OF THE PLANE OR 2-SPACE C D F E. NOTE THAT AS THE DIVIDED LINE (FIG 1) IS HIGHER SPACE TO THE DIVIDING POINT, SO THE PLANE IS HIGHER SPACE TO THE LINE THAT DIVIDES IT AND BOUNDS THE TWO ADJACENT PORTIONS

FIG 3.

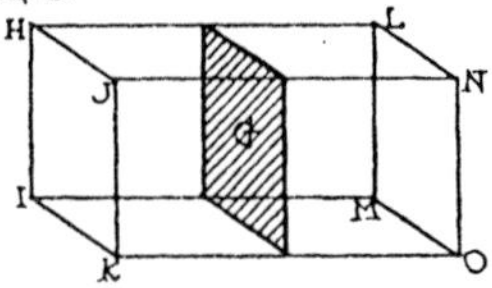

AGAIN: THE PLANE G (FIG 3), A 2-SPACE SEPARATES AND MUTUALLY BOUNDS TWO ADJACENT PORTIONS OF THE SOLID 3-SPACE H I K J N L M O

THEREFORE, WE HAVE AT LEAST RIGID ANALOGY TO JUSTIFY US IN SAYING THAT OUR 3-SPACE DIVIDES HYPERSPACE OF 4 DIMENSIONS, AND MUTUALLY BOUNDS TWO ADJACENT PORTIONS THEREOF.

THE POINT P IS WHOLLY CONTAINED IN THE LINE, EVERY POINT OF THE LINE A B IS IN CONTACT WITH ITS HIGHER SPACE, THE PLANE; EVERY PART OF THE PLANE G IS IMMERSED IN ITS HIGHER SPACE, THE SOLID TOUCHING EVERY POINT OF THE PLANE. SIMILARLY OUR 3-SPACE, AS A BOUNDARY IMMERSED IN 4-SPACE, MUST UNDENIABLY BE IN CONTACT AT EVERY POINT WITH THAT 4-SPACE THAT IS, THE INNERMOST PARTS OF OUR SOLIDS ARE AS OPEN TO TOUCH FROM 4-SPACE AS ARE ITS BOUNDARIES TO US.

IN GENERAL, ROTATION OCCURS IN 2-SPACE ABOUT A POINT, IN 3-SPACE ABOUT A LINE, AND IN 4-SPACE ABOUT A PLANE

FIG 1.

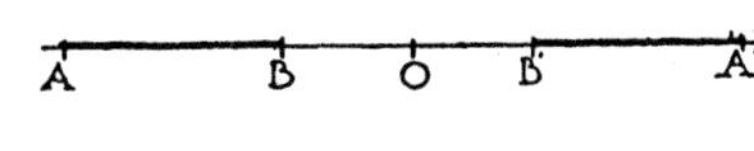

IF TWO LINES, A B AND A' B' IN THE SAME 1-SPACE ARE SYMMETRICAL IN RELATION TO A POINT O OF THAT SPACE (FIG. 1) A B CANNOT BE SO MOVED IN 1-SPACE THAT THE CORRESPONDING POINTS SHALL COINCIDE. TO EFFECT SUCH COINCIDENCE IT IS NECESSARY TO ROTATE A B THROUGH 2-SPACE ABOUT O AS A CENTER, OR, ROUGHLY SPEAKING, A B MUST BE TAKEN UP INTO 2-SPACE, TURNED OVER AND PUT DOWN ON B'A'.

FIG 2.

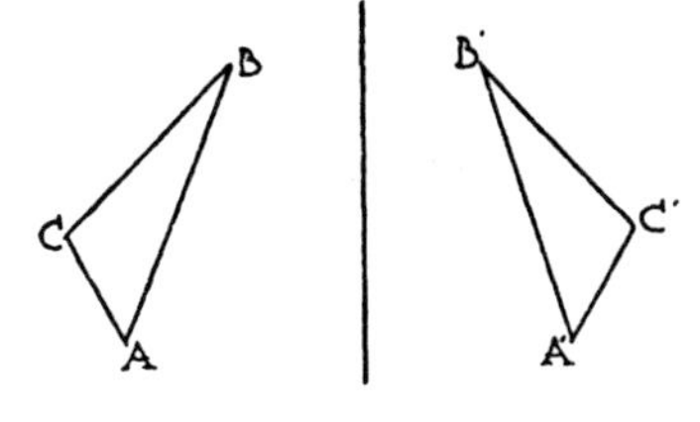

FIG 3

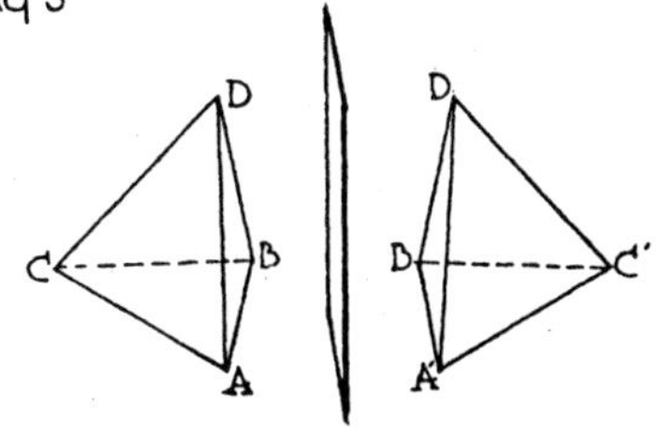

LIKEWISE, IF TWO TRIANGLES IN THE SAME 2-SPACE ARE SYMMETRICAL WITH RESPECT TO A LINE (FIG 2) COINCIDENCE OF CORRESPONDING POINTS AND LINES CAN BE EFFECTED ONLY BY ROTATING ONE TRIANGLE THROUGH 3-SPACE ABOUT THE AXIS OF SYMMETRY, THAT IS, ONE TRIANGLE MUST BE TAKEN UP INTO 3-SPACE, TURNED OVER, AND PUT DOWN ON THE OTHER

AGAIN, IF TWO POLYHEDRAL FIGURES IN THE SAME 3-SPACE ARE SYMMETRICAL WITH RESPECT TO A PLANE, BUT NOT WITH RESPECT TO ANY SINGLE LINE OR POINT (FIG 3), COINCIDENCE OF CORRESPONDING POINTS, LINES, AND PLANES CAN BE EFFECTED ONLY BY ROTATING ONE POLYHEDRAL FIGURE THRO' 4-SPACE ABOUT THAT PLANE, OR ROUGHLY SPEAKING, ONE POLYHEDRON MUST BE TAKEN UP INTO 4-SPACE, TURNED OVER, AND PUT DOWN ON THE OTHER. A MIRROR IMAGE OF A THING REPRESENTS THIS KIND OF A REVOLUTION, IMPOSSIBLE IN 3-SPACE.

SYMMETRY IN ANY SPACE THE EVIDENCE OF A HIGHER DIMENSIONAL ACTION

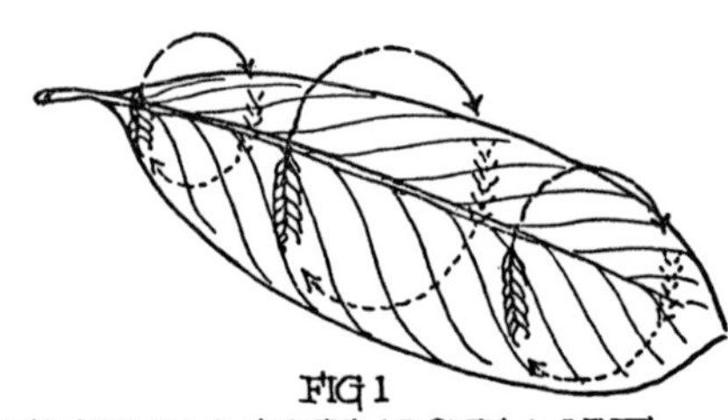

FIG 1
ROTATION IN 3 SPACE ABOUT A LINE

IT IS READILY CONCEIVABLE THAT THE 2-DIMENSIONAL SYMMETRY WHICH IS SO CONSTANT A CHARACTERISTIC OF VEGETABLE FORMS MAY BE THE RESULT OF A ROTATION ABOUT A CENTRAL AXIS, OF ETHERIC PARTICLES IN THE DIMENSION OF SPACE NOT CONTAINED WITHIN THE PLANE OF THE PETAL OR OF THE LEAF, I.E., IN ITS HIGHER SPACE

FIG 2
ROTATION IN 4 SPACE ABOUT A PLANE

MAY IT NOT BE THAT THE 3-DIMENSIONAL SYMMETRY WHICH IS SO UNIVERSALLY CHARACTERISTIC OF ANIMAL ORGANISMS IS THE RESULT OF BI-ROTATION, OR REVOLUTION ABOUT A PLANE: THE 4-DIMENSIONAL MOVEMENT ANALOGOUS TO THE TURNING ABOUT A LINE IN 3-SPACE? THE RIGHT AND THE LEFT HANDS, FOR EXAMPLE, MAY OWE THEIR CORRESPONDENCE OF PARTS TO A 4-DIMENSIONAL ROTATION IN THE MINUTE INVISIBLE MATTER OF OUR WORLD THE EFFECT OF THIS KIND OF A MOTION WOULD BE SUCH AS THE TWO HANDS SHOW: 3-DIMENSIONAL SOLIDS RELATED TO ONE ANOTHER AS OBJECT AND MIRROR IMAGE

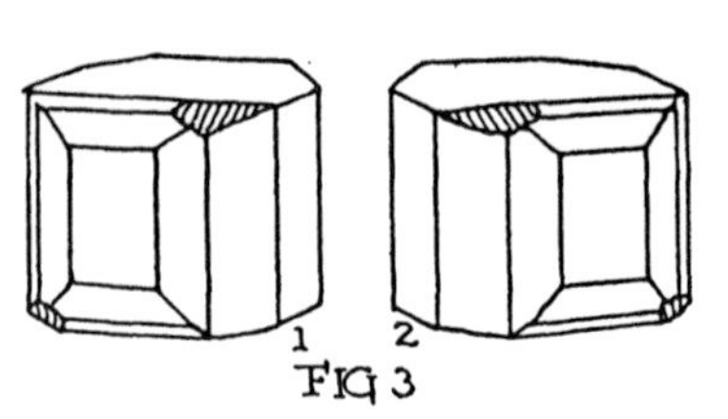

FIG 3

FIG 3 REPRESENTS CRYSTALS OF A TARTRATE BEARING THE RELATION OF OBJECT AND IMAGE

IF 1 CHANGED INTO 2 WITHOUT CHEMICAL RESOLUTION AND RECONSTITUTION IT WOULD INDICATE A 4TH DIMENSION

LOWER SPACES ARE CONTAINED IN HIGHER

(I.E., IN SPACES HAVING A LARGER NUMBER OF DIMENSIONS) AND THE FIGURES OF SUCH LOWER SPACES HAVE CLOSE RELATIONS TO CORRESPONDING FIGURES OF MORE DIMENSIONS. THUS ARE FOUND IN LOWER SPACES SIGNIFICANT SUGGESTIONS OF HIGHER.

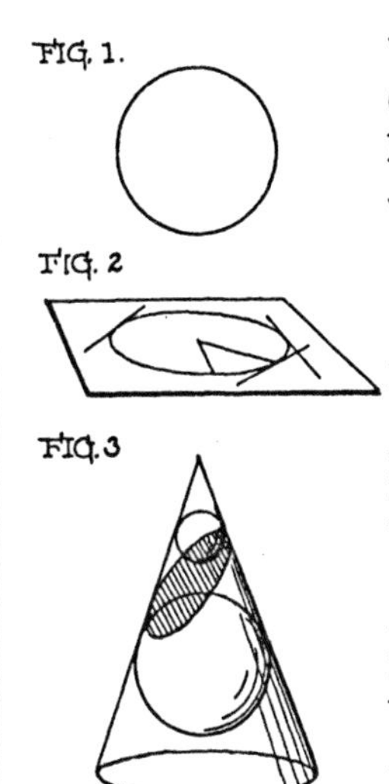

THE CIRCUMFERENCE OF A CIRCLE, CONSIDERED SIMPLY AS A LINE, AND AS POSSESSING BUT ONE DIMENSION, PRESENTS FEW RELATIONS. THE COMPLETE CIRCLE IN THE PLANE, HOWEVER (FIG 2), A 2-DIMENSIONAL FIGURE, HAS A RICH GEOMETRY OF ITS OWN, TREATING OF RADII, TANGENTS, CHORDS, ETC. IN CONNECTION WITH 3-SPACE THE CIRCLE HAS NUMEROUS ADDITIONAL AND IMPORTANT RELATIONS WITH THE SPHERE, CYLINDER, CONE, ETC. (FIG. 3). ANY OTHER FIGURE IN LIKE MANNER GAINS IMMENSELY IN INTEREST WHEN WE CONSIDER THE NEXUS OF ITS RELATIONS WITH HIGHER SPACE. FURTHERMORE, JUST AS CERTAIN LINEAR FIGURES REQUIRE 3-SPACE FOR THEIR GENERATION (THE HELIX, FIG. 4, FOR EXAMPLE), IN HYPERSPACE A WEALTH OF NEW AND INTERESTING LINES AND SURFACES ARE BROUGHT TO LIGHT.

THE FOLLOWING CONCRETE ILLUSTRATION MAY HELP TO SUMMARIZE THE FOREGOING IDEAS IN THE MIND: THE PATH OF A MAN MOVING ACROSS A HUGE SHEET OF ICE REPRESENTS ROUGHLY A LINE CONTAINED WITHIN A 2-SPACE, THE EARTH'S SURFACE. THIS SURFACE, BEING THE BOUNDARY OF THE EARTH'S SPHERE, REQUIRES, LIKE THE HELIX, THE SCOPE GIVEN BY 3-SPACE FOR ITS DESCRIPTION. — NOW, OWING TO THE TRAVEL OF THE SUN TOWARDS THE CONSTELLATION HERCULES THE EARTH REVOLVES ABOUT THE SUN NOT IN A PLANE, BUT IN A HELIX. MOREOVER, THE WHOLE STARRY UNIVERSE MAY BE SWEEPING ALONG A PATH OF WHICH THE 4TH DIMENSION IS A COMPONENT.

FOUR-DIMENSIONAL SPACE EXTENDS AWAY IN A DIRECTION AT RIGHT ANGLES TO ALL DIRECTIONS IN THREE-DIMENSIONAL SPACE, WHICH IS THEREFORE EXPOSED OR OPEN FROM THE REGION OF THE FOURTH DIMENSION

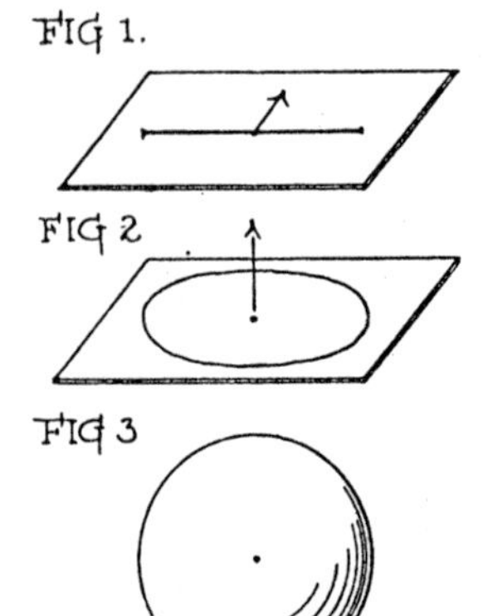

A POINT STARTING FROM THE CENTER OF A LINE (A 1-SPACE), AND MOVING OFF IN A PERPENDICULAR DIRECTION INTO A PLANE (A 2-SPACE) WILL NOT APPROACH ANY PORTION OF THE LINE. [FIG 1]

A POINT STARTING FROM THE CENTER OF A CIRCLE (A 2-SPACE) AND MOVING OFF ON A LINE PERPENDICULAR TO ITS PLANE WILL NOT APPROACH ANY PORTION OF THE CIRCUMFERENCE OF THE CIRCLE [FIG 2].

A POINT STARTING FROM THE CENTER OF A SPHERE [FIG 3] AND MOVING OFF ON A LINE PERPENDICULAR TO OUR SPACE WILL NOT APPROACH ANY PART OF THE SURFACE OF THE SPHERE, BUT WILL MOVE AWAY AT THE SAME RATE FROM ALL POINTS OF THIS SURFACE.

BECAUSE ANALOGY REQUIRES THAT THE BOUNDARIES OF HYPERSOLIDS (PORTIONS OF FOUR-DIMENSIONAL SPACE) BE THREE-DIMENSIONAL, IT THEREFORE FOLLOWS THAT EACH POINT WITHIN A SOLID IS ALL THAT IN ITS PLACE SEPARATES THE TWO PORTIONS INTO WHICH THE THREE-SPACE OF THE SOLID DIVIDES FOUR-DIMENSIONAL SPACE. EACH POINT IS THUS THE BEGINNING OF A PATHWAY INTO—AND OUT OF—FOUR DIMENSIONAL SPACE! DOES THE INDIVIDUAL, CLOTHED IN ITS FOUR-DIMENSIONAL VEHICLE FORSAKE THE PHYSICAL BODY AT DEATH BY THIS ROUTE? DOES AN OBCESSING ENTITY CAPTURE THE PHYSICAL BODY OF ANOTHER DURING THE TEMPORARY ABSENCE OF ITS RIGHTFUL TENANT BY ENTERING THROUGH THE UNGUARDED GATEWAY OF THE FOURTH DIMENSION OF SPACE?

EACH "SPACE," OR DIMENSIONAL ORDER, CONTAINS AN INFINITY OF SPACES OF DIMENSIONS FEWER BY ONE, AND IS ITSELF ONE OF AN INFINITE NUMBER OF SIMILAR SPACES CONTAINED WITHIN A SPACE OF DIMENSIONS GREATER BY ONE

FIG 1.

FIG 2.

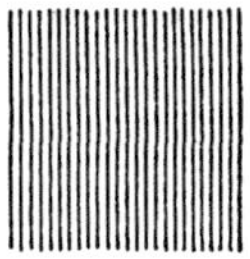

FIG 3.

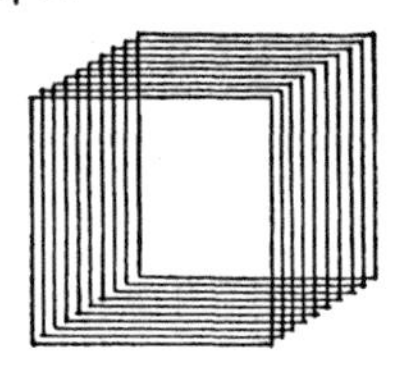

A LINE CONTAINS AN INFINITE NUMBER OF POINTS (FIG 1).

A PLANE CONTAINS AN INFINITE NUMBER OF LINES—1-SPACES (FIG 2).

A SOLID CONTAINS AN INFINITE NUMBER OF PLANES—2-SPACES (FIG 3).

BY ANALOGY, AN INFINITE NUMBER OF SOLIDS—3-SPACES—WOULD BE CONTAINED WITHIN A HYPER-SOLID—A 4-SPACE.

IF, AS PHILOSOPHERS AFFIRM, THE VISIBLE WORLD EXISTS ONLY IN, AND FOR CONSCIOUSNESS—IF IT IS BUT THE "PERCEPTION OF A PERCEIVER"—THEN FOR EACH CONSCIOUS PERSON THERE EXISTS A DIFFERENT WORLD.

IT FOLLOWS LOGICALLY THAT THESE COUNTLESS PERSONAL CONSCIOUSNESSES IN WHICH THE THREE-DIMENSIONAL PERCEPTION OF THE WORLD INHERES, MAY BE THOUGHT OF AS SO MANY 3-SPACES GOING TO FORM A HIGHER, OR FOUR-DIMENSIONAL UNITY—THE CONSCIOUSNESS OF HUMANITY AS A WHOLE. FOR IT IS CLEAR THAT HUMANITY IS HIGHER-DIMENSIONAL IN RELATION TO THE INDIVIDUAL MAN IF WE CONSIDER HUMANITY IN ITS TOTALITY, IT HAS POWERS OF WHICH NO SINGLE HUMAN BEING IS POSSESSED IT IS BOTH OLD AND YOUNG, YET DEATHLESS; IT IS IN ALL PLACES AT ONCE, IT SEES ALL OBJECTS, HEARS ALL SOUNDS, THINKS ALL THOUGHTS, EXPERIENCES ALL SUFFERINGS, ALL DELIGHTS. NOW SUPPOSE A MAN TO DWELL CONSTANTLY IN THE THOUGHT OF THIS HUMANITY, TO IDENTIFY ALL HIS INTERESTS WITH ITS LARGER INTERESTS, IS IT NOT THINKABLE THAT HE MIGHT TRANSCEND THE PERSONAL LIMITATION, AND MERGE HIMSELF INTO THE LARGER CONSCIOUSNESS OF WHICH HE HAS ALL THE WHILE BEEN A PART?

THE REPRESENTATION OF THE FORM OF AN OBJECT IS CONDITIONED AND RESTRICTED BY THE 'SPACE' IN WHICH SUCH REPRESENTATION OCCURS. THE HIGHER THE SPACE, THE MORE COMPLETE THE REPRESENTATION

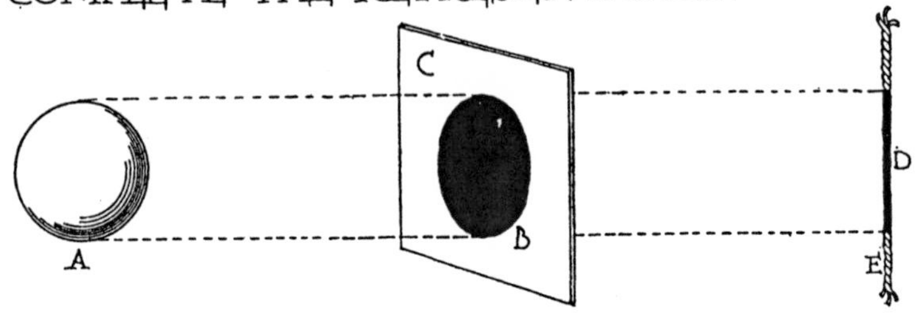

FOR EXAMPLE, THE SPHERE (A 3-SPACE FORM) CAN ONLY BE REPRESENTED IN PLANE SPACE BY A CIRCLE OF A DIAMETER EQUAL TO THE DIAMETER OF THE SPHERE, AND IN LINEAR SPACE BY A LINE OF A LENGTH EQUAL TO THE SAID DIAMETER.

THESE LOWER-DIMENSIONAL REPRESENTATIONS MAY BE CONCEIVED OF AS THE SHADOWS CAST BY HIGHER-SPACE FORMS ON LOWER SPACE WORLDS.

THE SPHERE A CASTS THE CIRCULAR SHADOW B UPON THE PLANE C, AND THE LINEAR SHADOW D UPON THE LINE E. OF WHAT, THEN, IS THE SPHERE ITSELF, IN THIS SENSE, THE SHADOW? THE HYPER-SPHERE: RELATED TO THE SPHERE AS IT IS RELATED TO ITS CIRCLE OF GREATEST DIAMETER.

PLATE 12

A NUMBER OF DISSIMILAR GEOMETRICAL FIGURES ARE SUSCEPTIBLE OF BEING CORRELATED AND COMBINED INTO A SINGLE HIGHER-DIMENSIONAL FORM

THIS FACT, AS REGARDS TWO, AND THREE DIMENSIONAL SPACE, IS WELL ILLUSTRATED BY MEANS OF THE FOLLOWING PUZZLE PROBLEM

QUESTION. WHAT FORM OF A WOODEN STOPPER WILL ANSWER EQUALLY FOR THREE HOLES IN A BOARD, ONE IN THE SHAPE OF A CIRCLE, ONE A SQUARE, AND ONE A TRIANGLE (FIG 1).

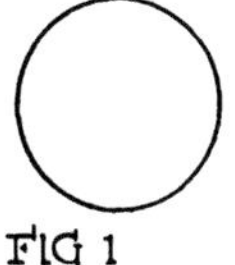

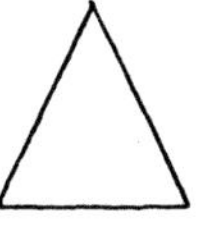

FIG 1

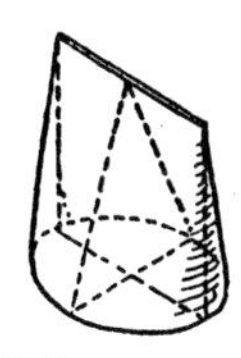

FIG 2.

ANSWER: A STOPPER WHICH WILL FILL ANY OF THE THREE HOLES IS A WEDGE-SHAPED SOLID HAVING A CIRCULAR BASE, ONE SQUARE SECTION, AND ONE TRIANGULAR (FIG 2)

BY ANALOGY, THE THREE-DIMENSIONAL CORRELATIVES OF THE CIRCLE, THE SQUARE, AND THE TRIANGLE, NAMELY, THE SPHERE, THE CUBE, AND THE TETRAHEDRON, MIGHT BE BOUNDARIES AND CROSS-SECTIONS OF SOME SINGLE FOUR-DIMENSIONAL FORM OF WHICH SUCH FIGURES AS THE CYLINDER, THE PRISM, AND THE PYRAMID MIGHT ALSO BE PROJECTIONS.

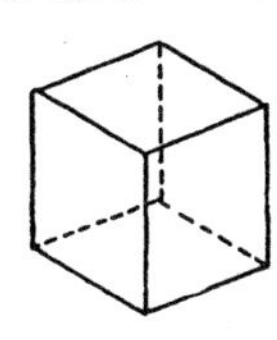
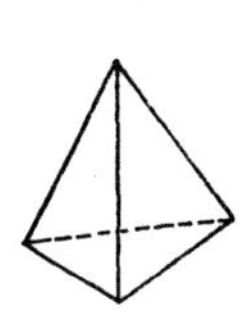
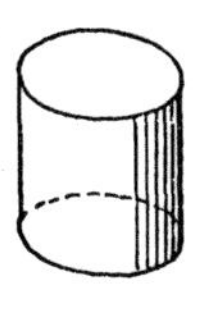
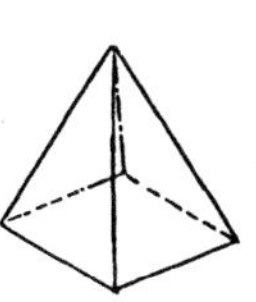
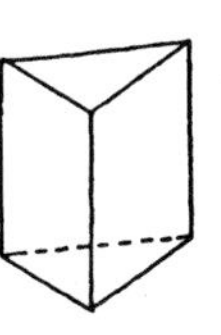

FIG 3

IF THE PERSONALITY IS THE PROJECTION, ON THE PHYSICAL PLANE, OF THE INDIVIDUALITY OR HIGHER SELF, A SINGLE INDIVIDUAL MIGHT PROJECT ITSELF IN MANY DIFFERENT PERSONALITIES, SEPARATED FROM ONE ANOTHER IN TIME

REINCARNATION MAY THUS BE CONCEIVED OF AS THE SUCCESSIVE REPRESENTATIONS OF A TRANSCENDENTAL SELF

PLATE 13

THE FOURTH DIMENSION MAY BE MANIFESTED TO US THROUGH CERTAIN MOTIONS IN OUR SPACE OF THREE DIMENSIONS BY TRANSLATING ITSELF FOR OUR EXPERIENCE, INTO TIME

FIG. 1

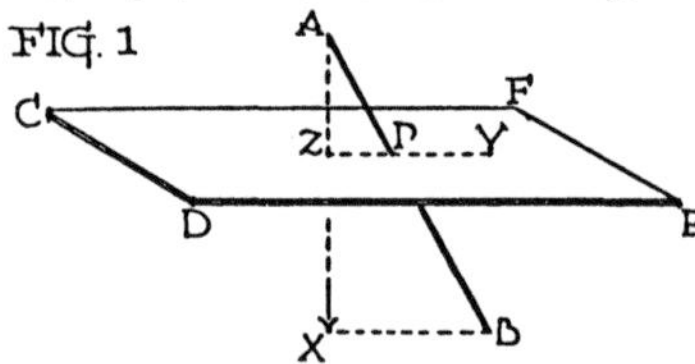

THIS IS BEST ILLUSTRATED BY CONSIDERING FIRST THE SAME TRUTH IN TERMS OF LOWER SPACE

THE LINE AB, HAVING EXTENSION IN THE DIMENSION NOT CONTAINED IN THE 2-SPACE CDEF, MOVES IN THIS ADDITIONAL THIRD DIMENSION; DOWNWARD, TO WIT. THE POINT P WHERE THE LINE PENETRATES THE PLANE, MOVES TO THE LEFT, FROM Y TO Z. AND THE TIME REQUIRED TO COMPLETE THIS MOTION FROM Y TO Z IN 2-SPACE MEASURES THE EXTENT PERPENDICULAR TO THE PLANE OF THE LINE AB IN THE UNCONTAINED DIMENSION. THAT IS, THE TIME MEASURES AX.

FIG. 2

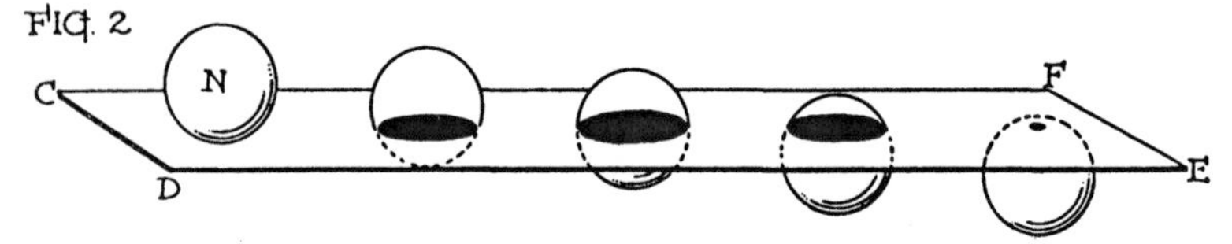

THE SPHERE N (FIG 2), IN TRANSIT ACROSS A 2-SPACE REPRESENTED BY THE PLANE CDEF, WOULD MANIFEST ITSELF IN THE PLANE AS A POINT, EXPANDING TO A CIRCLE WHICH WOULD ATTAIN A MAXIMUM DIAMETER EQUAL TO THAT OF THE SPHERE, AFTER WHICH IT WOULD SHRINK TO A POINT AND DISAPPEAR.

FIG 3

SIMILARLY A HYPERSPHERE OR 4-DIMENSIONAL SPHERE OF RADIUS R, PASSING THROUGH OUR SPACE, WOULD APPEAR TO US AS A SPHERE WITH RADIUS GRADUALLY INCREASING FROM ZERO TO R AND THEN GRADUALLY DECREASING FROM R TO ZERO (FIG 3)

THE PHENOMENA OF LIFE-GROWTH ARE SIGNIFICANTLY SUGGESTIVE OF 4TH DIMENSIONAL ENTRANCES UPON 3-SPACE EXPERIENCE

THE DENSITY OF BODIES AN INDICATION OF A PRESSURE FROM THE DIRECTION OF THE FOURTH DIMENSION AND A MEASURE OF EXTENSION IN THAT DIMENSION

FIG 1.

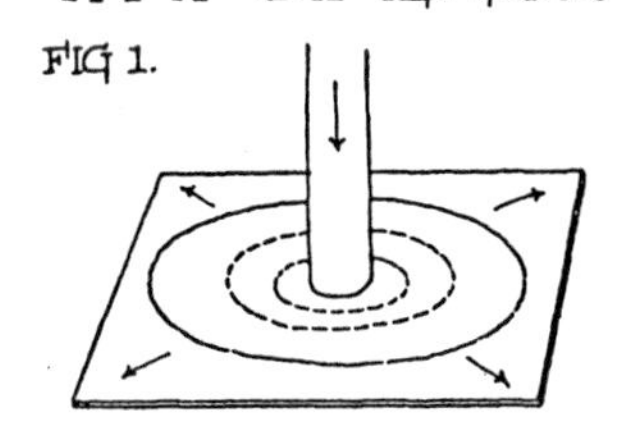

A STREAM OF WATER FALLING VERTICALLY UPON A PLANE SURFACE TENDS NATURALLY TO SPREAD OUT IN THE TWO DIMENSIONS OF THE PLANE, SETTING UP, IN SO DOING, UNDULATIONS IN THE SHAPE OF ENLARGING CONCENTRIC CIRCLES, DIMINISHING AS TO DEPTH. THE RAPIDITY OF THIS LATERAL EXTENSION, AND THE FORCE AND HEIGHT OF THE WAVES WILL DEPEND UPON THE HEIGHT FROM WHICH THE STREAM OF WATER FALLS, THAT IS TO SAY, UPON ITS PRESSURE IN THE THIRD, OR VERTICAL, DIRECTION. [FIG 1]

CARRYING OUT THE ANALOGY, IN OUR WORLD OF THREE DIMENSIONS THE EXPANSIVE FORCE OF GASES WOULD BE DUE TO SOME SIMILAR INFLUX FROM THE REGION OF THE FOURTH DIMENSION, AND THE AMOUNT OF PRESSURE EXERTED BY A GAS WOULD BE A MEASURE OF FOUR-DIMENSIONAL EXTENSION. SO LONG AS THE QUANTITY OF ENERGY COMING DOWN FROM A HIGHER WORLD IS NOT EXPENDED, THERE WOULD BE SOME DEGREE OF FORCE ENTERING BY WAY OF THE FOURTH DIMENSION WHICH CAUSES THE GAS TO DILATE IN OUR THREE-DIMENSIONAL WORLD. THE CAPACITY OF A GAS TO EXPAND COMES THUS FROM A FOUR-DIMENSIONAL WORLD.

THE DENSITY OF SOLID BODIES WOULD BE DUE TO THE SAME CAUSE, WITH THIS DIFFERENCE, THAT THEY ARE STABLE, AND CANNOT DILATE; THAT IS TO SAY, THEY ARE IN EQUILIBRIUM WITH ATMOSPHERIC PRESSURE. AS A CONSEQUENCE, THE VARIATIONS IN THE DENSITIES OF BODIES WOULD BE DUE TO VARIATIONS IN THE FORCE EXERTED FROM THE FOURTH DIMENSION.

THE FOURTH DIMENSION CAN THUS BE CONSIDERED AS REPRESNTED BY THE DENSITY OF SOLIDS, OR BY THE EXPANSIVE FORCE OF GASES.

EVOLUTION INTERPRETED IN TERMS OF HIGHER SPACE

IF WE PASS A HELIX (A SPIRAL IN THREE DIMENSIONS), THROUGH A FILM (A 2-SPACE), THE INTERSECTION WILL GIVE A POINT MOVING IN A CIRCLE [A, FIG 1]. IF THE HELIX REMAINS STILL AND THE FILM MOVES VERTICALLY UPWARD, THE SPIRAL WILL BE REPRESENTED IN THE FILM BY THE CONSECUTIVE POSITIONS OF THE POINT OF INTERSECTION. IN THE FILM THE PERMANENT EXISTENCE OF THE SPIRAL WILL BE EXPERIENCED AS A TIME SERIES—THE RECORD OF TRAVERSING THE SPIRAL WILL BE A POINT MOVING IN A CIRCLE.

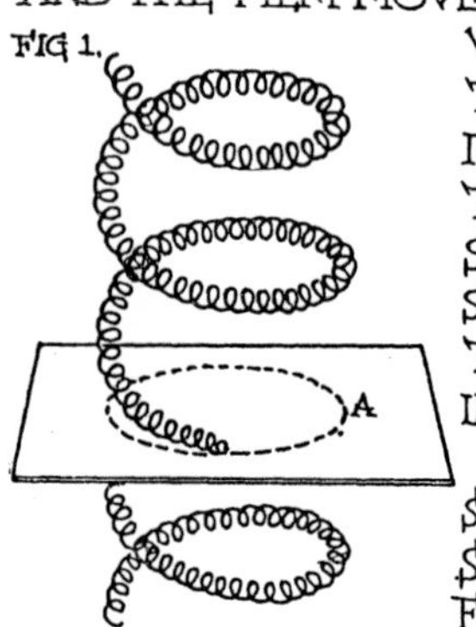

IT IS EASY TO IMAGINE COMPLICATED STRUCTURES OF THE NATURE OF THE SPIRAL, STRUCTURES CONSISTING OF FILAMENTS, AND TO SUPPOSE ALSO THAT THESE STRUCTURES ARE DISTINGUISHABLE FROM EACH OTHER AT EVERY SECTION. IF WE CONSIDER THE INTERSECTIONS OF THESE FILAMENTS WITH THE FILM AS IT PASSES TO BE THE ATOMS CONSTITUTING A FILMAR UNIVERSE, WE SHALL HAVE IN THE FILM A WORLD OF APPARENT MOTION; WE SHALL HAVE BODIES CORRESPONDING TO THE FILAMENTARY STRUCTURE, AND THE POSITIONS OF THESE STRUCTURES WITH REGARD TO ONE ANOTHER WILL GIVE RISE TO BODIES IN THE FILM MOVING AMONGST ONE ANOTHER. THIS MUTUAL MOTION IS APPARENT MERELY. THE REALITY IS OF PERMANENT STRUCTURES STATIONARY, AND ALL THE RELATIVE MOTIONS ARE ACCOUNTED FOR BY ONE STEADY MOVEMENT OF THE FILM.

NOW IMAGINE A <u>FOUR</u>-DIMENSIONAL SPIRAL PASSING THROUGH A <u>THREE</u>-DIMENSIONAL SPACE. THE POINT OF INTERSECTION, INSTEAD OF MOVING IN A CIRCLE, WILL TRACE OUT A SPHERE. ASSUMING, AS BEFORE, A COMPLICATED STRUCTURE FOR THE SPIRAL, ITS PRESENTMENT IN 3-SPACE WILL CONSIST OF BODIES BUILT UP OF SPHERES OF VARIOUS MAGNITUDES MOVING HARMONIOUSLY AMONGST ONE ANOTHER, AND REQUIRING <u>TIME</u> FOR THEIR DEVELOPMENT. MAY NOT THE ATOM, THE MOLECULE, THE CELL,—THE EARTH ITSELF, BE SO MANY PATHS AND PATTERNS OF AN UNCHANGING UNITY?

LOWER SPACE SYSTEMS IN OUR WORLD

HIGHER SPACE SPECULATION IS LARGELY BASED ON THE ANALOGY SUPPLIED BY A HYPOTHECATED 2-SPACE CONSCIOUSNESS LIMITED TO THE TWO DIMENSIONS OF A PLANE, OR BY A 1-SPACE CONSCIOUSNESS CONFINED TO A LINE. HAVE SUCH CONDITIONS OF EXISTENCE ANY REALITY?—FOR IF THEY HAVE NOT IT IS USELESS TO ATTEMPT TO PROVE BY THEM THE REALITY OF A 4-DIMENSIONAL WORLD. IF, AS SEEMS PROBABLE, ALL LIFE IS CONSCIOUSNESS, AS MUCH "AWARE" AS ITS SPATIAL LIMITATIONS PERMIT, NATURE EVERYWHERE ABOUNDS IN SUCH LOWER SPACE SYSTEMS: THEY STAND EXHIBITED, INDEED, IN NEARLY EVERY PLANT THAT GROWS, FOR

"THE POINT, THE LINE, THE SURFACE, AND THE SPHERE
IN SEED, STEM, LEAF, AND FRUIT APPEAR."

A TOMATO PLANT EXEMPLIFYING A SYSTEM OF LINES IN ITS STALK AND STEMS, A SYSTEM OF PLANES IN ITS LEAVES, AND A SYSTEM OF SOLIDS IN THE FRUIT, IN OTHER WORDS, A 1, 2, AND A 3 SPACE WORLD.

ALTHOUGH SUCH LINES AND PLANES ARE REALLY 3-DIMENSIONAL, THEY YET FULFIL THE CONDITIONS AND CONFORM TO THE DEFINITION OF LOWER SPACES, WHICH DO NOT PRECLUDE EXTENSION IN HIGHER DIMENSIONS, SO LONG AS THERE IS LITTLE OR NO FREEDOM OF MOVEMENT THEREIN. IN STALKS AND STEMS THE FREE MOVEMENT OF PARTICLES IS RESTRICTED IN ALL DIRECTIONS SAVE ONE. IN LEAVES IT IS RESTRICTED IN THE THIRD DIMENSION

SNOW CRYSTAL

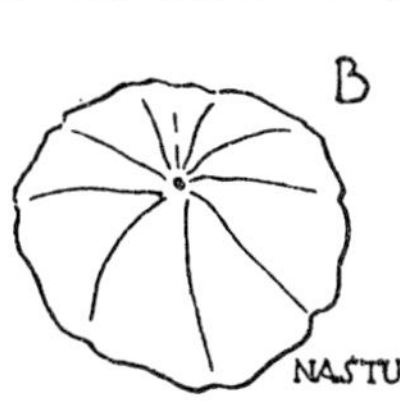

NASTURTIUM LEAF

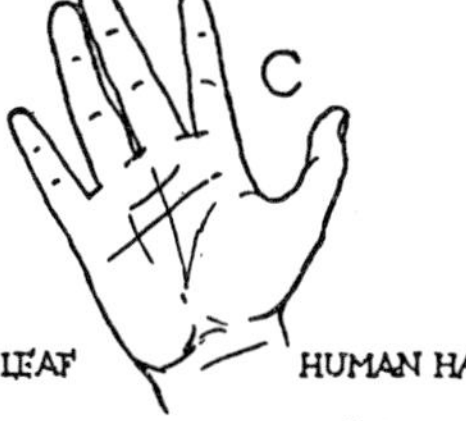

HUMAN HAND

A CONSCIOUSNESS CONTAINED IN SUCH A FORM AS "A" WOULD IN EFFECT INHABIT A 1-SPACE, "B" A 2-SPACE, AND "C" A 3-SPACE.

AN INTERPRETATION OF CERTAIN SO-CALLED PSYCHIC PHENOMENA IN TERMS OF THE HIGHER-SPACE THEORY

APPARITION IN A CLOSED SPACE

FIG 1 FIG 2

FIG 1 REPRESENTS WHAT WOULD CORRESPOND, IN 2-SPACE, TO A ONE-ROOM HOUSE TO ENTER IT WITHOUT PASSING THROUGH ANY OF THE BREAKS IN THE PERIMETER REPRESENTING DOOR AND WINDOWS WOULD CONSTITUTE A "PSYCHIC" PHENOMENON IN 2-SPACE, BUT IT COULD BE ENTERED BY WAY OF 3-SPACE BY A REVOLUTION ABOUT THE LINE OF ONE OF ITS SIDES. SIMILARLY, A 3-SPACE HOUSE (FIG 2), MIGHT BE ENTERED BY A 4-DIMENSIONAL ROTATION ABOUT THE PLANE OF ONE OF ITS SIDES

POSSESSION, OBSESSION, AUTOMATIC WRITING, AND ALLIED PHENOMENA ARE SUSCEPTIBLE OF EXPLANATION BY MEANS OF THE HIGHER-SPACE HYPOTHESIS. IT IS ONLY NECESSARY TO REALIZE THAT FROM THE HIGHER REGION OF SPACE THE INTERIOR OF A SOLID IS AS EXPOSED AS THE INSIDE OF A PLANE FIGURE IS EXPOSED FROM THE REGION OF THE THIRD DIMENSION—THE HEART COULD BE PLUCKED FROM THE BODY WITHOUT BREAKING THE SKIN. AN ALIEN INVADING WILL, CLOTHED IN MATTER CAPABLE OF A FOUR-DIMENSIONAL ROTATION, MIGHT THUS POSSESS ITSELF OF THE HAND OR BRAIN

CLAIRVOYANCE IN SPACE

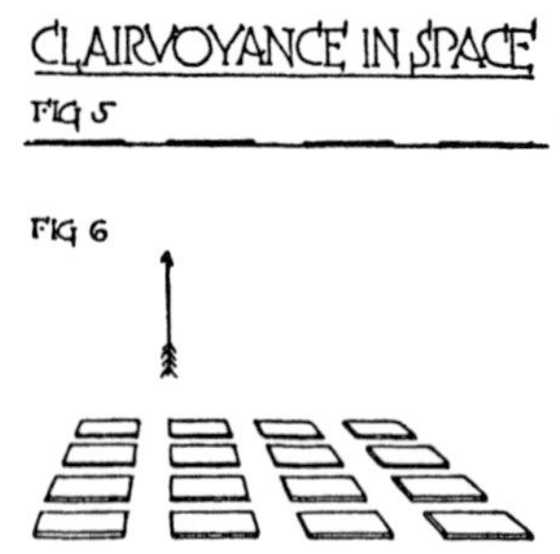

IMAGINE A CITY IN 2-SPACE WITH HOUSES SIMILAR TO FIG 1. TO NORMAL 2-SPACE PERCEPTION THEY WOULD APPEAR AS IN FIG 5, ONLY THE LINES FORMING THE WALLS VISIBLE, THE NEARER ONES CONCEALING ALL THOSE MORE REMOTE. BY RISING IN THE THIRD DIMENSION THE FIELD OF VISION WOULD ENLARGE SO AS TO INCLUDE THE ENTIRE CITY OF HOUSES. THE INTERIORS WOULD THEN BECOME VISIBLE, THOUGH OBLIQUELY. FROM HIGH OVERHEAD THE INTERIORS WOULD OPEN BROADLY INTO VIEW. CORRESPONDINGLY, BY MOVING OUT OF 3-SPACE INTO 4-SPACE, THE FIELD OF PERCEPTION WOULD EXPAND CUBICALLY AND THE INTERIORS OF ALL 3-SPACE SOLIDS WOULD STAND REVEALED BEING EXPOSED IN THE REGION OF THE 4TH DIMENSION.

A 2-SPACE "MAN" INHABITING A PLANE WOULD SEE ONLY THE LINES BOUNDING THE "SOLIDS" (PLANE FIGURES) OF HIS WORLD

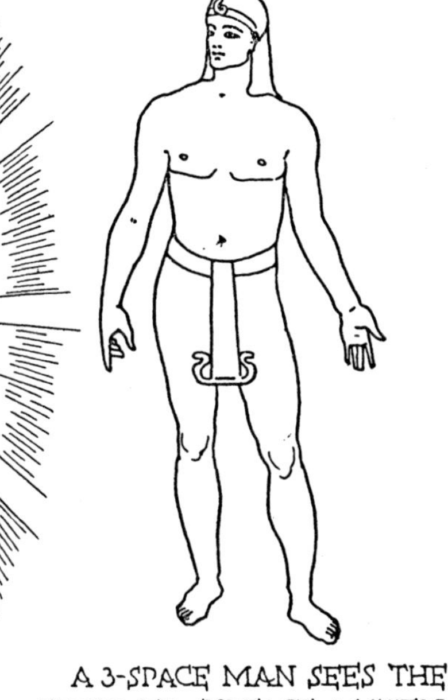

A 3-SPACE MAN SEES THE ENCLOSED SURFACE AS WELL AS THE BOUNDARIES OF SUCH 2-SPACE "SOLIDS", PERCEIVING THEM TO BE NOT REALLY SOLIDS, BUT BOUNDARIES OR CROSS-SECTIONS OF THE SOLIDS OF HIS WORLD—THE THINGS WHICH HE KNOWS TO BE 3DIMENSIONAL, BUT OF WHICH HE CAN SEE ONLY THE OUTSIDES —— BY ANALOGY, FROM A 4TH DIMENSION THESE SAME SOLIDS WOULD IN TURN APPEAR TRANSPARENT AND BE PERCEIVED TO BE BUT BOUNDARIES OR CROSS-SECTIONS OF 4-DIMENSIONAL SOLIDS —— CLAIRVOYANT VISION IS OF THIS ORDER, INDICATING THAT IT IS 4-DIMENSIONAL. SEEN CLAIRVOYANTLY, THE INTERNAL STRUCTURE OF THE HUMAN BODY IS VISIBLE WITHIN ITS CASING, ALSO THE AURA, OR HIGHER-DIMENSIONAL BODY

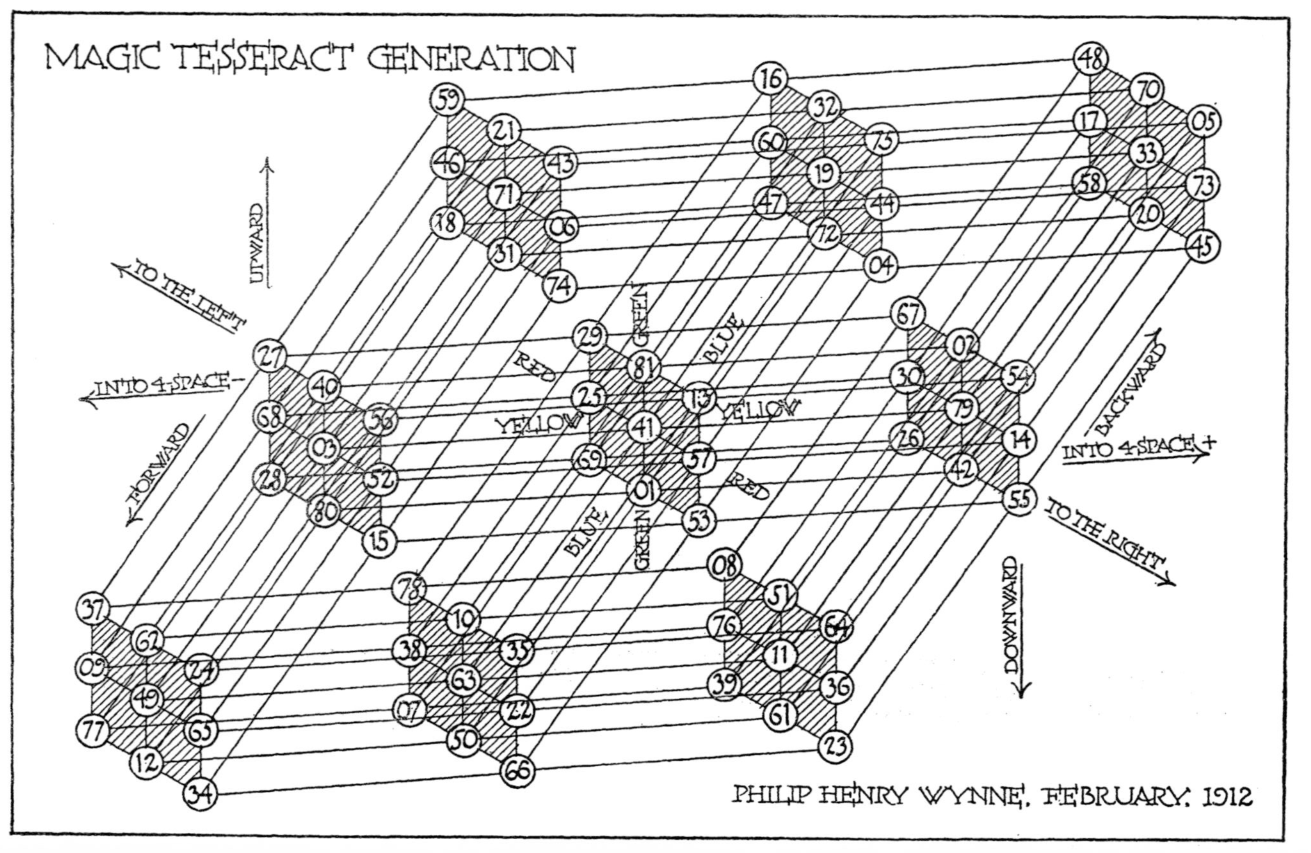
MAGIC TESSERACT GENERATION
UPWARD
TO THE LEFT
INTO 4-SPACE −
FORWARD
RED
YELLOW
GREEN
BLUE
BACKWARD
INTO 4-SPACE +
TO THE RIGHT
DOWNWARD
PHILIP HENRY WYNNE. FEBRUARY, 1912

PLATE 20

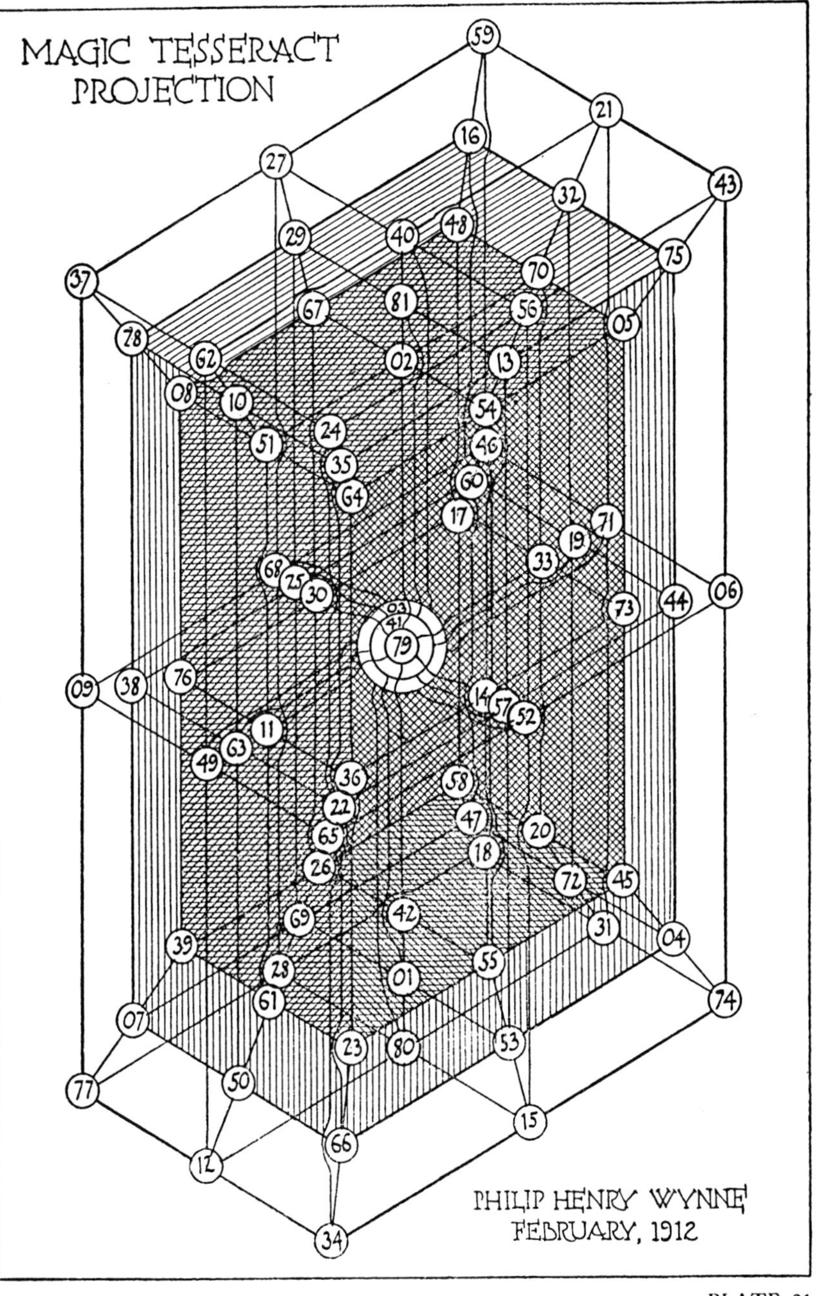
MAGIC TESSERACT
PROJECTION
59
21
16
27
43
32
48
29
40
75
70
37
67
81
56
05
78
62
02
13
08
10
54
24
51
46
35
60
64
17
71
19
68
25
33
30
06
44
73
03
41
79
09
38
76
14
57
52
11
63
49
36
58
22
47
20
65
18
26
72
45
42
69
31
04
39
28
55
01
74
61
07
23
80
53
50
77
15
66
12
34
PHILIP HENRY WYNNE
FEBRUARY, 1912

PLATE 21

INTRODUCTION TO THE MAGIC TESSERACT.

THE UNTRAINED MIND FINDS SOMETHING ESPECIALLY SATISFACTORY AND SECURE IN NUMERICAL RELATIONS. IT EASILY BELIEVES THAT WHAT IS TRUE OF 4, 17, OR 1001 UNITS MUST BE TRUE OF 4, 17, OR 1001 APPLES—MILES, OR MEN. THE AUTHOR ACCORDINGLY INVOKED THE MATHEMATICAL GENIUS OF PHILIP HENRY WYNNE TO FURNISH HIM WITH SOME INEXPUGNABLE ILLUSTRATION, FOUNDED UPON THE PROPERTIES OF NUMBER, OF AN OPEN ROAD FOR HUMAN THOUGHT INTO THE FOURTH DIMENSION.

THE RESULT OF MR. WYNNE'S DIVING INTO THE DEEP WATERS OF MATHEMATICS WAS THE PRODUCTION OF THE MAGIC TESSERACT, A PEARL WHICH ANYBODY IS FREE TO EXAMINE AND ADMIRE, BUT ONE WHICH ONLY A MATHEMATICIAN CAN PROPERLY APPRAISE. IF THE READER WILL FOLLOW THE ENSUING EXPLANATION STEP BY STEP, VERIFYING THE RELATIONS NOTED, HE WILL BE ABLE TO PARTICIPATE, WITHOUT DEEP KNOWLEDGE OR HARD LABOR, IN THE ASSURANCE OF THE UNDERLYING REALITY OF HYPERSPACE WHICH COMES TO THE MATHEMATICIAN OF AN OPEN MIND, AS A RESULT OF HIS RESEARCHES.

1	14	15	4
8	11	10	5
12	7	6	9
13	2	3	16

FIG 1.

A MAGIC SQUARE OF FOUR

THE ACCOMPANYING FIGURE REPRESENTS A MAGIC SQUARE OF 4 MADE BY A WELL KNOWN METHOD. THE READER SHOULD VERIFY THE FOLLOWING RELATIONS, FOR THEY ARE NOT TRIVIAL IN CONNECTION WITH WHAT COMES LATER.

EACH HORIZONTAL AND EACH VERTICAL COLUMN ADDS 34. EACH DIAGONAL ADDS 34. FOUR CORNER CELLS ADD 34. FOUR CENTRAL CELLS ADD 34. TWO MIDDLE CELLS OF TOP ROW ADD 34 WITH TWO OF BOTTOM ROW; SIMILARLY WITH MIDDLE CELLS OF RIGHT AND LEFT COLUMNS. GO ROUND THE SQUARE CLOCKWISE: 1ST CELL BEYOND 1ST CORNER + 1ST BEYOND 2D + . . . 3RD + . . . 4TH = 34. EACH CORNER SET OFF BY HEAVY LINES ADDS 34.

TAKE ANY NUMBER AT RANDOM; FIND THE THREE OTHER NUMBERS CORRESPONDING TO IT IN ANY MANNER WHICH RESPECTS SYMMETRICALLY TWO DIMENSIONS AND THE SUM OF THE FOUR NUMBERS WILL BE 34.

A MAGIC CUBE OF FOUR

THIS DRAWING REPRESENTS A MAGIC <u>CUBE</u> OF FOUR CONSTRUCTED BY A PURE EXTENSION OF THE <u>SQUARE</u> METHOD OF FIG 1. PLATE 22.

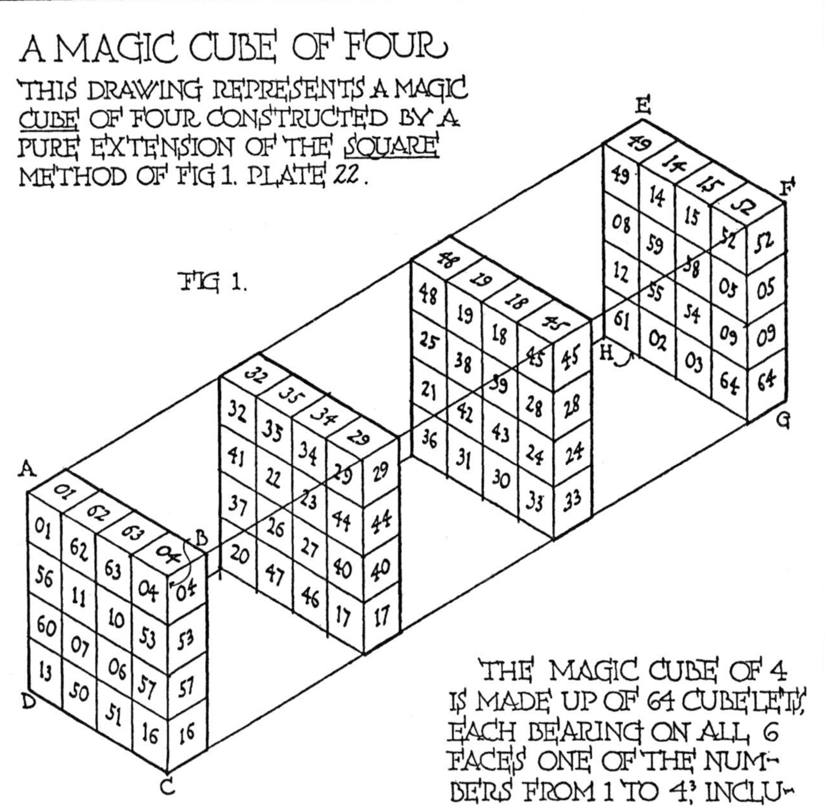

THE MAGIC CUBE OF 4 IS MADE UP OF 64 CUBELETS, EACH BEARING ON ALL 6 FACES ONE OF THE NUMBERS FROM 1 TO 4^3 INCLUSIVE. NOW THE CUBE CAN BE SLICED INTO 4 VERTICAL SECTIONS FROM LEFT TO RIGHT AS IN FIG 1. WHICH SHOWS THE SECTIONS SEPARATED SO THAT THE INTERIOR SUMMATIONS CAN BE SEEN.

OR IT CAN BE SEPARATED INTO OTHER 4 VERTICAL SECTIONS BY CUTTING PLANES PERPENDICULAR TO THE EDGE A B—PROCEEDING FROM FRONT TO BACK.

OR THE 4 SECTIONS MAY BE HORIZONTAL, MADE BY PLANES PERPENDICULAR TO A D.

NOW EACH ONE OF THESE 12 SECTIONS PRESENTS A MAGIC <u>SQUARE</u> IN WHICH EACH ROW AND EACH COLUMN ADDS 130. THE DIAGONALS OF THESE SQUARES DO <u>NOT</u> ADD 130, BUT THE 4 DIAGONALS OF THE CUBE <u>DO</u> ADD 130

THE READER SHOULD VERIFY A FEW OF THESE SUMMATIONS IN EACH OF THE THREE SETS OF SECTIONS.

A FURTHER CONSIDERATION OF MAGIC FIGURES

A DETAILED STUDY OF THE MAGIC CUBE OF 4 WILL SERVE TO CONVINCE THE READER OF THE ESSENTIALLY SOLID NATURE OF THE SPACE ARRANGEMENT AND THE PERFECT CONTINUITY OF THE "MAGICAL" PROPERTIES OF NUMBER FROM TWO DIMENSIONS TO THREE.

NOW THESE SAME NUMERICAL PROPERTIES CONTINUE UNINTERRUPTEDLY INTO THE FOURTH DIMENSION. TO FIX THE IDEA MORE CONCRETELY, CONSIDER THE FOLLOWING STATEMENTS BASED ON THE POWERS OF THE NUMBER 7.

FIG 1.

7 2 3 4 5 6 1

FIG 2.

30	39	48	1	10	19	28
38	47	7	9	18	27	29
46	6	8	17	26	35	37
5	14	16	25	34	36	45
13	15	24	33	42	44	4
21	23	32	41	43	3	12
22	31	40	49	2	11	20

FIG 1. REPRESENTS A MAGIC LINE OF 7. OBSERVE THAT THE SUMS OF PAIRS OF NUMBERS EQUIDISTANT FROM 4 IS

$$\frac{7^1+1}{2} \times 2$$

FIG 2 REPRESENTS A MAGIC SQUARE OF 7. OBSERVE THAT HERE THE VARIOUS SUMMATIONS OF 7 NUMBERS GIVE

$$\frac{7^2+1}{2} \times 7$$

IN THE MAGIC CUBE OF 7 THE SUMMATIONS ARE

$$\frac{7^3+1}{2} \times 7$$

IN THE MAGIC TESSERACT THE SUMMATIONS ARE

$$\frac{7^4+1}{2} \times 7$$

THE SUM IN THE CASE OF THE MAGIC LINE IS NOT ANOMALOUS IN THAT $\frac{7+1}{2}$ IS MULTIPLIED BY 2 INSTEAD OF BY 7; IT IS DUE TO THE FACT THAT WE ARE ADDING BUT 2 NUMBERS TOGETHER INSTEAD OF 7 NUMBERS AS IN THE OTHER CASES. IF WE TOOK ALL 7 NUMBERS OF THE MAGIC LINE WE SHOULD HAVE ONLY ONE SUM AND NO "MAGICAL" CORRESPONDENCES.

NOT ONLY SUCH SERIES AS 1, 2, 3, 4 BUT ARITHMETICAL PROGRESSIONS IN GENERAL, GEOMETRICAL PROGRESSIONS, AND OTHER SERIAL FUNCTIONS SUBMIT TO MAGICAL ARRANGEMENTS IN N DIMENSIONS.

A MAGIC SQUARE OF THREE—A CUBE OF THREE

OBSERVE THAT A MAGIC LINE CANNOT BE FORMED OF LESS THAN 4 NUMBERS, OR A MAGIC SQUARE OF LESS THAN 9 [FIG 1].

FIG 1.

8	1	6
3	5	7
4	9	2

A MAGIC CUBE REQUIRES AT LEAST 27 NUMBERS [FIG 2]. AND EVEN WITH THIS NUMBER THERE ARE MANY LIMITATIONS DUE TO LACK OF SCOPE FOR NUMEROUS COMBINATIONS POSSIBLE WITH CUBES OF 5, OR BETTER, OF 7.

MAGIC CUBE OF THREE

EACH ROW AND EACH COLUMN OF EACH OF THE 9 MAGIC SQUARES (3 SQUARES TO EACH DIMENSION) ADDS 42.

EACH OF THE 4 CUBE DIAGONALS ADDS 42.

EACH OF THE 6 DIAGONALS OF THE 3 MAGIC SQUARES CONTAINING THE CENTRAL NUMBER 14, ADDS 42

FIG 2.

ANY RANDOM LINE DRAWN THROUGH THE CENTER OF THE CUBE WILL CUT TWO SURFACE CELLS WHOSE NUMBERS WILL ADD TWICE THE CENTRAL NUMBER 14. ALL THESE LINES ARE "MAGIC" LINES.*

OUR WELL KNOWN NUMBERS 1, 2, 3, 4 ETC. CONTAIN AMONG THEM FOUR-DIMENSIONAL MAGIC PATHS AS REAL AND DEMONSTRABLE AS THOSE IN 2-SPACE.

A MAN WILL OFTEN GIVE "TWO AND TWO MAKES FOUR" AS AN EXAMPLE OF PERFECT CERTITUDE. THESE MAGIC NUMBER PROPERTIES RETORT ON HIM BY CHALLENGING HIS DENIAL OF THE FOURTH DIMENSION WHICH HIS OWN WELL-TRUSTED NUMBERS AFFIRM MOST INSISTENTLY

---

* THIS CUBE OF 3 (AN ODD NUMBER) HAS A CENTRAL CUBELET, WHICH GIVES RISE TO SEVERAL INTERESTING PROPERTIES LACKING IN THE 4-CUBE.

EXPLANATION OF THE MAGIC TESSERACT

THE ATTEMPT WILL NOW BE MADE TO RENDER CLEARLY INTELLIGIBLE MR. WYNNE'S MAGIC TESSERACT THE READER IS URGED TO VERIFY FOR HIMSELF THE SUMMATIONS PARTICULARLY SPECIFIED AND THE VARIOUS OTHER RELATIONS POINTED OUT. A DETAILED STUDY OF THE FIGURE WILL RESULT IN THE UNFOLDING OF SOME OF THE AMAZING AND BEAUTIFUL INTER-RELATIONS THAT LIE OUTSIDE 3-SPACE HORIZONS.

FIRST WE MUST CONSIDER THE SYMBOLS AND CONVENTIONS USED. INSTEAD OF THE USUAL MAGIC SQUARE DIAGRAM OF FIG 1 THE ARRANGEMENT OF FIG. 2 WILL BE EMPLOYED.

8	1	6
3	5	7
4	9	2

FIG 1.

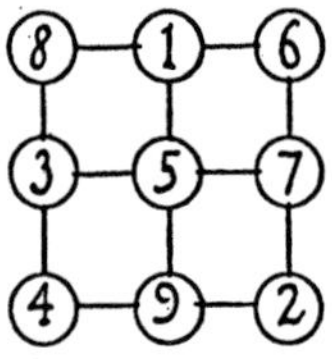

(8)	(1)	(6)
(3)	(5)	(7)
(4)	(9)	(2)

FIG 2.

IN FIG. 2 THE NUMBERS MAY BE IMAGINED TO BE ENCLOSED IN CRYSTAL SPHERES SUPPORTED ON A WIRE FRAMEWORK WHICH ALLOWS US TO SEE INTO THE INTERSPACES. THE WIRES MAY BE BENT ASIDE IF NECESSARY, AND ARE SUPPOSED TO BE CAPABLE OF EXTENSION OR SHORTENING AT OUR PLEASURE, THOUGH NOT SPONTANEOUSLY ELASTIC. PLATE 20 REPRESENTS 6 RECTANGULAR PRISMS STRETCHED OUT FROM ORIGINALLY CUBICAL FORM BY THE EXTENSION OF THE BLUE AND THE YELLOW WIRES. THIS DISTORTION IS OF COURSE TO EXHIBIT THE NUMBERS MORE CLEARLY, AND THE PRISMS SHOULD BE CONCEIVED AS PUSHED BACK AFTER INSPECTION INTO THEIR PROPER CUBICAL FORMS.

REMEMBER, THEN, THAT RED, GREEN, BLUE, YELLOW WIRES ARE NORMALLY OF EXACTLY EQUAL LENGTH.

FURTHER (SEE PLATE 20):—

(R)ED STANDS FOR MOTION TO THE (R)IGHT;

(G)REEN " " " TOWARDS THE (G)ROUND (DOWNWARDS);

(B)LUE " " " (B)ACKWARDS (AWAY FROM US);

(Y)ELLOW " " " IN THE POSITIVE SENSE OF THE FOURTH DIMENSION; INTO (Y)AMAPURA.

THE MAGIC TESSERACT—CONTINUED

NOW TURN TO PLATE 20—CONSIDER THE SPHERE CONTAINING 37; LET IT MOVE TO THE (R)IGHT ALONG THE (R)ED LINE, LEAVING AT EQUAL INTERVALS LIKE SPHERES FOR 62 AND 24. THERE ARE THEN 3 SPHERES, THIS BEING A 3-TESSERACT. OBSERVE THAT 37 + 62 + 24 = 123, THE MAGIC SUM: $= [(1+81) \div 2] \times 3$, OR, MORE GENERALLY, $\Sigma = \frac{N+N^5}{2}$, THE <u>CORRECT SUM FOR 4-SPACE</u>. [FIG 1].

FIG 1.

NEXT THE LINE 37, 62, 24, SHALL MOVE DOWNWARD—TOWARDS THE (G)ROUND—ALONG (G)REEN LINES, LEAVING 2 SIMILAR LINES, TO WIT, 9, 49, 65, 77, 12, 34. EACH LINE YIELDS THE MAGIC SUM 123, SO ALSO DO THE 3 VERTICAL LINES. WE THEREFORE HAVE A <u>MAGIC 3-SQUARE</u> [FIG 2].

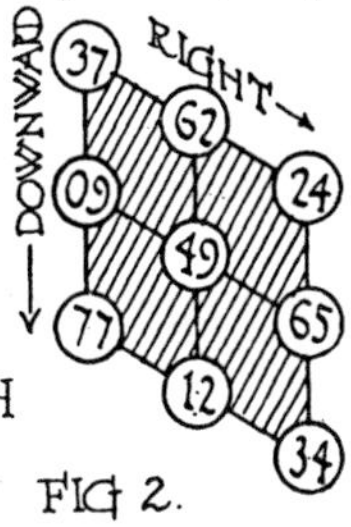

THIS SQUARE SHALL NOW MOVE (B)ACKWARDS ALONG (B)LUE PATHS, LEAVING TWO SIMILAR SQUARES AND GENERATING THE FIRST (STRETCHED-OUT) <u>MAGIC 3-CUBE</u> WHICH WE WILL CALL 37—74. NOTE THAT ALL THE BLUE LINES AS WELL AS BOTH THE OLD AND THE NEWLY-GENERATED RED AND GREEN LINES GIVE THE MAGIC SUM 123. [FIG 3].

FIG 2.

THE DIAGONALS OF THE MAGIC SQUARES AND OF THE MAGIC CUBES DO <u>NOT</u> ADD 123 EXCEPT WHEN THEY CROSS THE CENTER OF THE <u>TESSERACT</u>. ALL 8 OF THE TESSERACT DIAGONALS <u>DO</u> ADD 123. BOTH THESE FACTS ARE AS THEY SHOULD BE THOUGH THE HIGHER RELATIONS WHICH ARE THE CAUSES NEED NOT BE EXPOUNDED HERE.

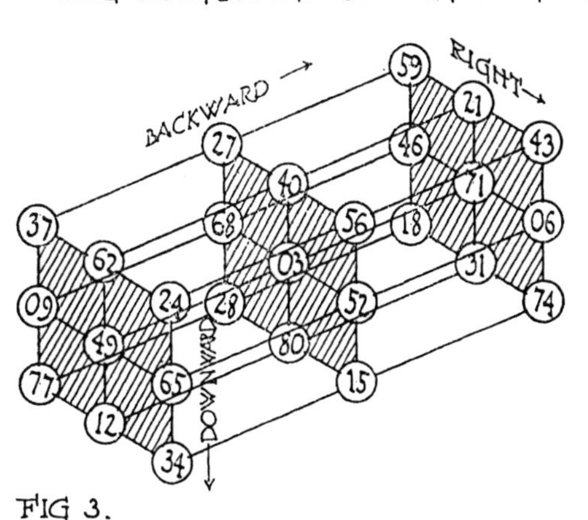

FIG 3.

THE MAGIC TESSERACT — CONTINUED —

NOW WE WILL TAKE OUR LONG LEAP OUT INTO THE DARK.

LET OUR FIRST CUBE, 37–74, MOVE OFF PERPENDICULARLY IN THE + SENSE OF THE FOURTH DIMENSION, LEAVING IN 4-SPACE TWO SIMILAR 3-SPACE CUBES, AND GENERATING THE MAGIC TESSERACT. THE PATHS OF THE CRYSTAL SPHERES ARE (SYMBOLICALLY) REPRESENTED BY THE SYSTEM OF LINES DESIGNATED AS YELLOW.

IF WE REALLY HAVE ACHIEVED THE MAGIC TESSERACT, THEN EVERY RED (MARKED) WIRE (1), GREEN WIRE (2), BLUE WIRE (3), YELLOW WIRE (4), MUST BEAR 3 NUMBERS YIELDING THE MAGIC SUM 123. THIS WILL BE FOUND TO BE THE CASE.

OBSERVE THAT ALL 4 DIMENSIONS ARE PERFECTLY EQUIPOTENT. FOR CLEARNESS THE WIRES MARKED BLUE AND YELLOW HAVE BEEN STRETCHED, BUT THIS MIGHT HAVE BEEN DONE TO THE RED AND TO THE GREEN INSTEAD. WHEN THE TESSERACT HAS BEEN PUSHED BACK INTO UNDISTORTED FORM THE RED, GREEN, BLUE, YELLOW, WIRES ARE ALL OF EQUAL LENGTH, AND EACH OF THE 108 WIRES CARRIES 3 NUMBERS WHOSE SUM IS 123.

THIS BEING SO, WHO SHALL SAY WHICH DIMENSION IS MORE REAL THAN THE OTHERS, AND WHY?

THUS IS THE MAGIC TESSERACT "GENERATED." LET US NOW SEE IF WE CAN CONFIRM OUR SOMEWHAT ANALYTIC CONSIDERATION BY COLLECTING OUR RESULTS IN A MORE IMAGINABLE FORM. IN SHORT, LET US TRY TO PROJECT BACK FROM 4-SPACE THE CONCRETE ASSEMBLAGE OF NUMBERS WHICH CONSTITUTE THE MAGIC TESSERACT.

WE FIND THAT WE CAN. THE RESULT IS SHOWN IN PLATE 21.

THE TINTING WILL SUGGEST TO THE EYE AN OUTER AND AN INNER CUBE, WHICH ARE RESPECTIVELY 37–74 AND 08-45 OF PLATE 20 (A AND C, FIG 1). BUT THE READER MUST ALSO IMAGINE ANOTHER CUBE, 78-04 OF PLATE 21 (B, FIG 1), HALF WAY BETWEEN THE OTHER TWO (SEE 78, 35, 66, 07, 47, 16, 75, 04, IN THE PROJECTION ALSO).

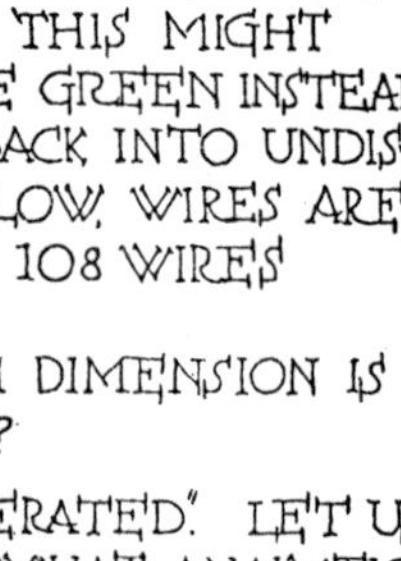

A
B
C

FIG 1.

THE MAGIC TESSERACT—CONCLUDED

THE READER SHOULD SATISFY HIMSELF BY TRIAL THAT EVERY LINE IN PLATE 20 HAS ITS CORRESPONDING LINE IN PLATE 21. ALSO THAT EACH LINE OF THE 108 IN THE PROJECTION CARRIES THE SAME 3 NUMBERS AS THE CORRESPONDING LINE OF THE 108 IN THE GENERATION.

THE FOLLOWING POINTS ARE IMPORTANT. IT IS PURELY ARBITRARY WHICH CUBE IS PUT APPARENTLY "WITHIN" OR "WITHOUT." THEY ARE REALLY SIDE BY SIDE IN 4-SPACE AND MERELY APPEAR TO BE CONTAINED ONE WITHIN ANOTHER; EXACTLY AS IN A PERSPECTIVE SKETCH OF A 3-SPACE ROOM INTERIOR THE PARALLELOGRAM OF THE FURTHER WALL IS APPARENTLY CONTAINED WITHIN THAT OF THE NEARER. FIG 1 ILLUSTRATES THIS AND ALSO ANOTHER IMPORTANT POINT, VIZ. THAT THE APPARENTLY SLANTING LINES AND PLANES OF THE TESSERACT ARE PURELY AN ILLUSION OF PERSPECTIVE. IN FIG 1, A F, B G, ETC ARE PART OF A SYSTEM OF 3 SETS OF PERPENDICULARS, THOUGH THEY APPEAR TO SLANT, WHILE A B APPEARS PERPENDICULAR TO A D, ETC. FURTHER, OBSERVE THAT IT WAS THE EXIGENCIES OF REPRESENTATION THAT COMPELLED, IN THE PROJECTION, THE PLACING OF SPHERE 79 INSIDE OF 41, AND THAT INSIDE OF 03—THEY REALLY FALL ONE BEHIND ANOTHER IN 4-SPACE.

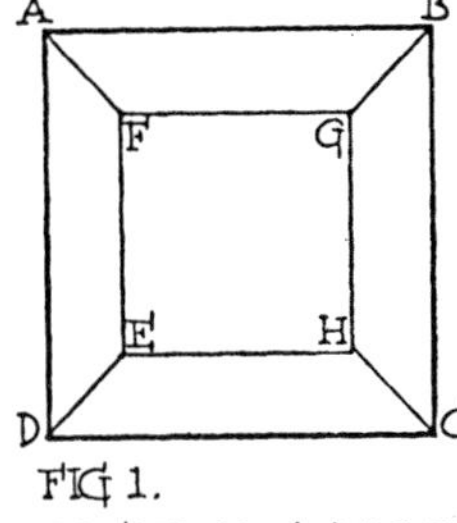

FIG 1.

IT IS INTERESTING AT THIS POINT TO IDENTIFY A FEW OF THE TESSERACT DIAGONALS AND TO VERIFY THEIR SUMMATIONS. JUST AS IN FIG. 1 A DIAGONAL EXTENDS FROM ONE CORNER OF THE OUTER SQUARE ACROSS THE CENTRAL, TO THE OPPOSITE CORNER OF THE INNER, SO IN THE TESSERACT WE PASS FROM THE OUTERMOST CUBE CORNER TO THE OPPOSITE CORNER OF THE INNERMOST CUBE, CROSSING AND INCLUDING THE TESSERACT CENTER, SPHERE 41 THUS;—37+41+45=123; 18+41+64=123 ETC.

THE FOREGOING WILL SUFFICE TO EXPLAIN THE SKETCHES AND ENABLE THE READER TO EXAMINE THE CURIOUS ASSEMBLAGE OF CONCEPTS AND RELATIONS WHICH CONSTITUTE THE MAGIC TESSERACT.

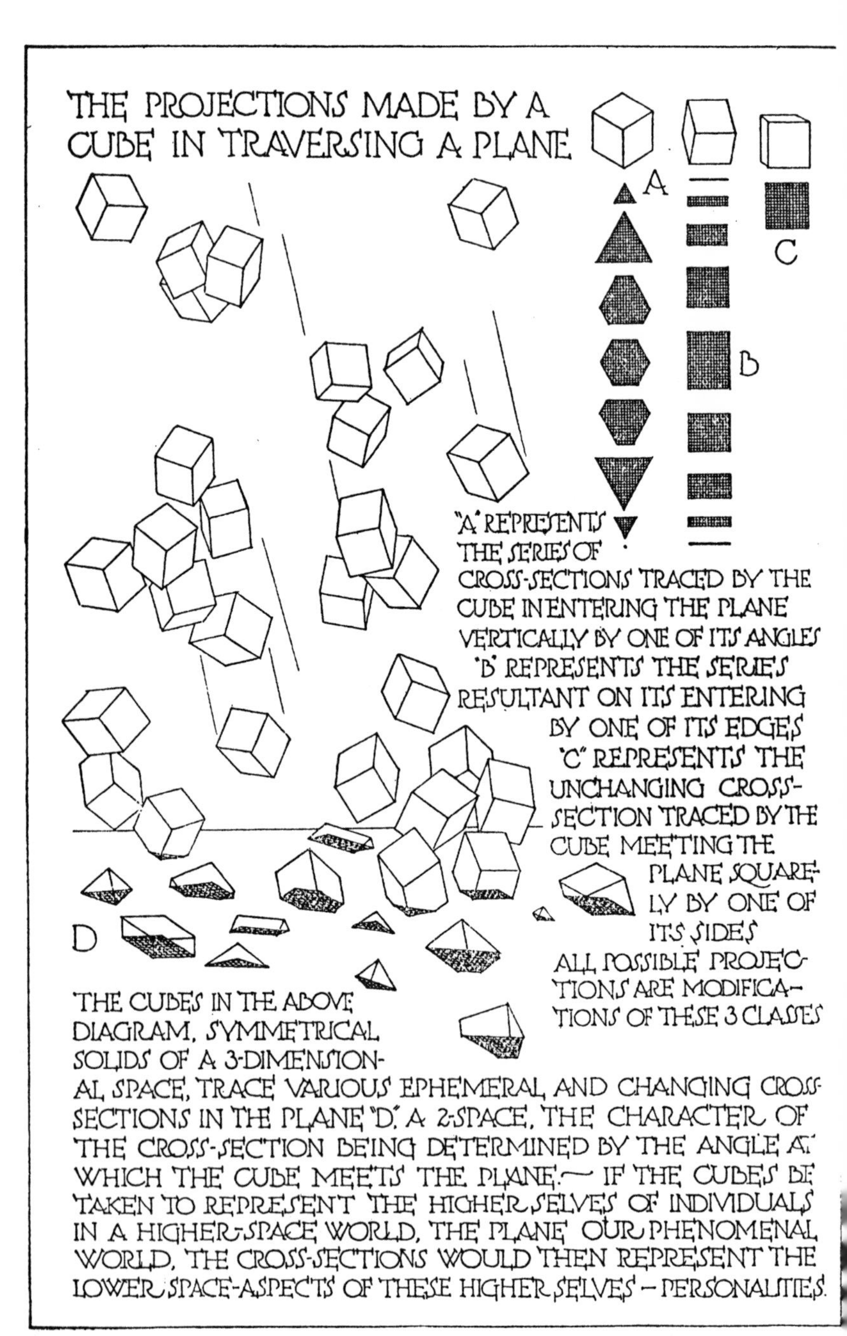
THE PROJECTIONS MADE BY A CUBE IN TRAVERSING A PLANE
A
B
C
D
"A" REPRESENTS THE SERIES OF CROSS-SECTIONS TRACED BY THE CUBE IN ENTERING THE PLANE VERTICALLY BY ONE OF ITS ANGLES
"B" REPRESENTS THE SERIES RESULTANT ON ITS ENTERING BY ONE OF ITS EDGES
"C" REPRESENTS THE UNCHANGING CROSS-SECTION TRACED BY THE CUBE MEETING THE PLANE SQUARELY BY ONE OF ITS SIDES
ALL POSSIBLE PROJECTIONS ARE MODIFICATIONS OF THESE 3 CLASSES
THE CUBES IN THE ABOVE DIAGRAM, SYMMETRICAL SOLIDS OF A 3-DIMENSIONAL SPACE, TRACE VARIOUS EPHEMERAL AND CHANGING CROSS-SECTIONS IN THE PLANE "D," A 2-SPACE, THE CHARACTER OF THE CROSS-SECTION BEING DETERMINED BY THE ANGLE AT WHICH THE CUBE MEETS THE PLANE.— IF THE CUBES BE TAKEN TO REPRESENT THE HIGHER SELVES OF INDIVIDUALS IN A HIGHER-SPACE WORLD, THE PLANE OUR PHENOMENAL WORLD, THE CROSS-SECTIONS WOULD THEN REPRESENT THE LOWER SPACE-ASPECTS OF THESE HIGHER SELVES — PERSONALITIES.

PLATE 30

MAN THE SQUARE

A HIGHER SPACE PARABLE

"Artful nature has given to the most perfect animal the same six limits as the cube has, most perfectly marked. . . . Man himself is, as it were, a cube."

—*Mysterium Cosmographicum.*
Kepler.

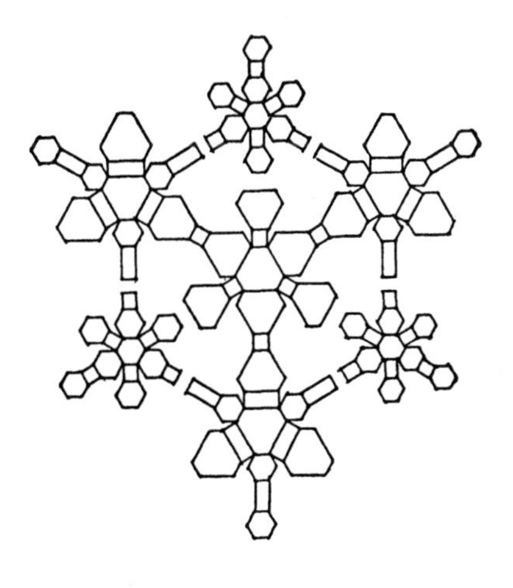

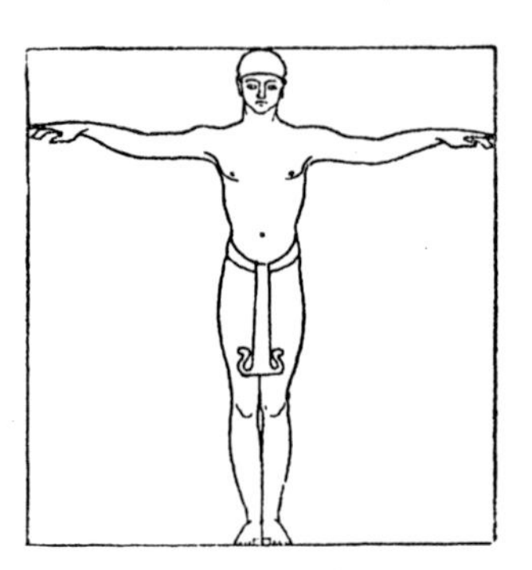

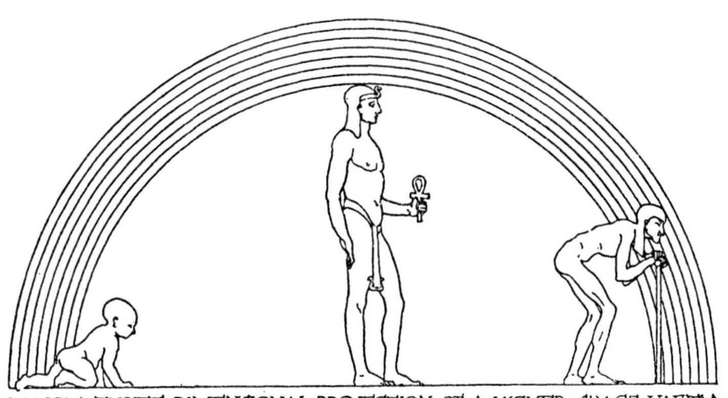

MAN: A THREE-DIMENSIONAL PROJECTION OF A HIGHER-SPACE UNITY

MAN THE SQUARE

THE CUBE AND THE SQUARE

"The phenomenal world receives its culmination and reflex of all in MAN. Therefore he is the mystic square—in his metaphysical aspect—the Tetraktys; and becomes the cube on the creative plane."

The Secret Doctrine. Vol. II, p. 39, Third Edition.

H. P. BLAVATSKY.

"And the city [the New Jerusalem] lieth four square, and the length is as large as the breadth. The length and the breadth and the height of it are equal."

The Revelation of St. John the Divine, xxi:16.

HERE ARE two examples of that order of recondite mystical truths so remote from ordinary knowledge and experience, or else so clothed with symbolism, as to be unintelligible to all save the initiated.

We are told of a certain correspondence or identity between the phenomenal world and man, and the symbolization of the two by the square. We are further instructed that on the "creative plane"—presumably some higher world of causes—the square is, or becomes, a cube. This idea, embodied in the first quotation, is borne out by the statement in the second, that the New Jerusalem, the dwelling place of perfected humanity (which might be a world or might be a body) has "the length and the breadth and the height of it" equal—in other words, is a cube.

What meaning lies here concealed? In order to discover it, let us try the experiment of taking the quoted statements not figuratively—as they were doubtless intended to be taken—but literally, and see where we are then able to come out.

With the simple-mindedness and confidence of children, let us picture to ourselves the phenomenal world, not as symbolically, but actually, a square. The square should not be conceived of as a purely geometric plane, i. e., having no thickness in the third dimension, for in that case it would have no physical existence. It should be thought of rather as an almost infinitesimally thin film of matter separating two portions of the cube—its higher space world—from each other: in other words, a cross-section of the cube. Fix this image clearly in mind: the creative plane, that is, the archetypal world, or world of causes, a crystal cube "like unto clear glass," divided midway by an iridescent film, the phenomenal world, made up of matter in a different state or condition—analogous, let us say, to oil on water, or to the bubble which sometimes appears in the neck of a bottle. [Illustration 1].

MACROCOSM AND MICROCOSM

This crystal cube would then represent the macrocosm, of which man—archetypal man—is the microcosm. *"As is the great, so is the small: as is the outer, so is the inner."* We must conceive of the great cube as containing many small cubes, replicas of itself in everything except size. *"Nothing is great: nothing is small."* This difference in size between the great cube and the small ones need not disturb us: it has no importance, for if the small cubes are conceived of as themselves containing still smaller cubes after the manner in which they are contained within the great cube, then every part of each has its correspondence in the other, and is capable of being expressed by the same ratios. If the great cube were shrunk to the size of one of the small cubes, there would be no difference whatever between them. If, on the other hand, the small cubes were expanded to the size of the great cube, the same would be true. This relativity of space-magnitude —a difficult thing to understand by minds untrained in philosophy and metaphysics—is apprehended unconsciously by everyone in such a concrete exemplification of it as is afforded by photography, for example. At a moving picture exhibition we never think of the image on the screen as being essentially

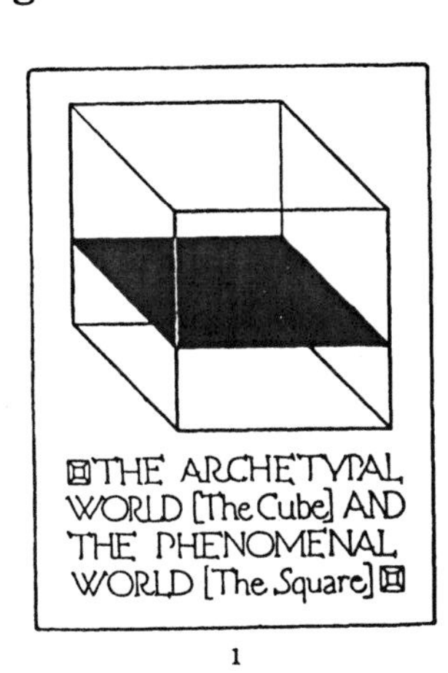
THE ARCHETYPAL WORLD [The Cube] AND THE PHENOMENAL WORLD [The Square]

1

different from the photograph from which it is projected, though the difference in spatial extension is enormous. All that matters is the relation of parts to one another, and these being identical, the question of absolute size does not even enter the mind.

The small cubes should be thought of as moving about within the limits of the great cube, such motion bringing them repeatedly in contact with the filmy plane which corresponds to the phenomenal world. They would register their passage though the matter of this film world by tracing in it countless cross-sections of themselves. If, as has been assumed, these small cubes correspond to the higher, or divine selves of men, identical in form and substance with the Great Self, their "Father in Heaven," then each transit of each cube, or individual, through the film square would be for it a physical incarnation, and the correlated succession of cross-sections which it traced in any one transit would represent one phenomenal life. Attributing consciousness to the individual cube—to its totality and to its every part—that aspect of consciousness stirred from latency to activity by contact with the matter of the film world in passing through it during one phenomenal life, would constitute the *personal* consciousness.

BIRTH AND DEATH

Each personality, each new projection of the cube in the plane, would be "born," so to speak, with the cube's initial contact with the film—since only thus and then could personal consciousness arise—and each would disappear or "die" with the final contact. The stream of impressions would be, as ours is, linear, i. e., successive—a constant becoming. All things would seem to be vanishing irrevocably into the void of time. There would be no survival, no immortality, for the personal consciousness on its own plane of manifestation, since the film matter which gave it, for the moment, form, would flow together and shape itself into new and different figures, the cross-sections of other cubes—new personalities.

With the cube—the true individual—the case would be different: the cube would know itself not to be "born" nor to "die." Each of its "personalities," or the tracings which it made in passing through the film, would inhere within it, since every conceivable cross-section is embraced within the cube itself. It would neither gain nor lose them by passing through the film: it would only manifest them in the matter of a lower spatial world.

The cube consciousness (individual) would embrace all cross-sectional consciousness (personal): it would have full knowledge of the film world, since it would apprehend that world from a higher region of space; but the cross-sectional or plane consciousness—a fragment, as it were, of the cube consciousness—would depend for its knowledge of the things of its world upon the constantly shifting line bounding the plane figure traced in the film by the cube in passing through. This line would be its vehicle for sensation. What report would such a vehicle make to the indwelling personal consciousness: what notion would that consciousness get, through this channel, of its world and the things of its world—and of the higher world?

FILM PHENOMENA

In order to answer this question at all adequately, it will be necessary to know something more of "film phenomena," to go into a more detailed analysis of the transit of solids through a plane. Picture, if you please, these thousands of little cubes streaming, so to speak, through a plane, meeting it at every possible angle and tracing cross-sections of themselves in transit, in the same way that the surface of a liquid traces the successive cross-sections of any solid introduced into it, as the fluid separates and flows together again. A moment's reflection will make plain the fact that the cubes, although identical in shape and size, would create plane figures widely divergent from one another, the differences being caused by the variation of angle at which each individual cube happened to encounter the plane. [Illustration 2]. If, for example, a cube entered by one of its corners, with its longest internal diagonal perpendicular to the plane, the first presentment of it would be a point, the meeting place of three adjoining faces. The trihedral section bounded by these faces would then trace itself out as an expanding equilateral triangle, until the three lowest corners of the cube became involved, when this triangle would change, by reason of the truncation of each apex, into a hexagon with three long sides and three short ones. The long sides would grow shorter and the short sides longer, as the cube continued its descent: there would be a moment when all six sides were of equal length, after which the forms would succeed each other in an inverse order, the hexagon changing into a triangle, which would shrink to a point and disappear, as the cube passed beyond the limits of the plane. If the cube should enter by one of its edges, the first presentment of it would be a line, which would thicken, so to speak, into a parallelogram, whose long sides, in the ensuing motion, would remain constant, and whose short sides would lengthen until they exceeded the other two by the amount that a parallelogram expressing the diagonal plane of the cube is longer than one of its square faces. This maximum attained, the section would shrink again, the changes occurring, as before, in an inverse order; ending, as they began, in a line, and an evanishment. Only in case the cube should happen to meet the plane squarely by one of its faces—that is, perpendicularly to the plane—would the resulting cross-section undergo no change throughout the period of transit. In this case, the tracing would be a square. [Illustration 3. See also plate 30].

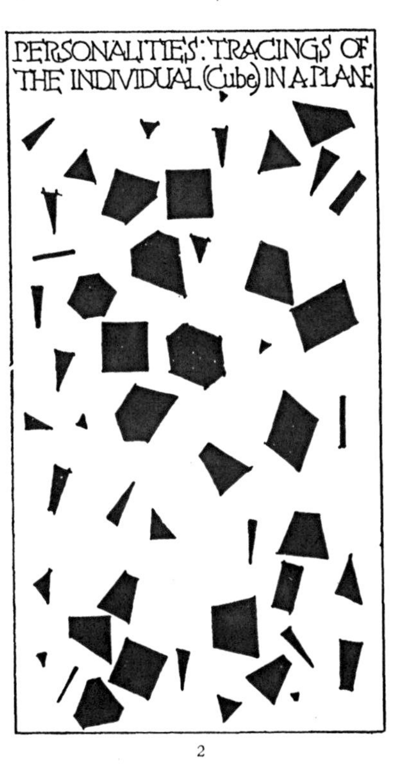

2

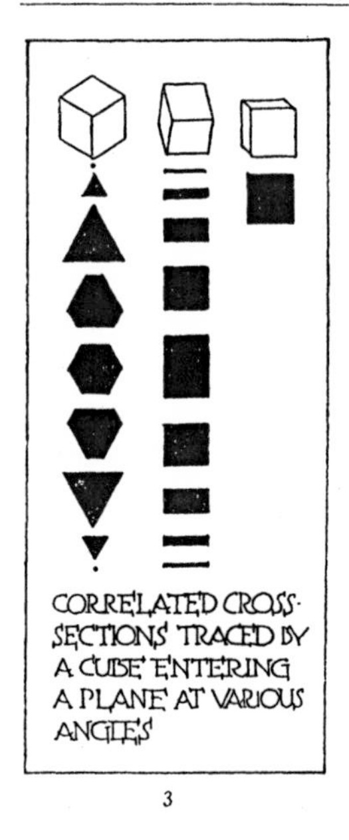

CORRELATED CROSS-SECTIONS TRACED BY A CUBE ENTERING A PLANE AT VARIOUS ANGLES

3

Now the angle at which the cube could meet the plane might vary almost infinitely, and each variation would result in a changing cross-section of different form, though every one of these would be referable to one or the other of the three general types described above. If the cubes descended, not vertically, but obliquely, the plane presentments of them, besides waxing and waning and changing their outline, would have, in addition, a lateral motion, great or small in proportion to the obliquity of the cubes' orbits; and thence, too, would arise a difference as to the figures in their duration, the more nearly vertical the descent, the shorter the period of phenomenality.

Bearing clearly in mind that each one of these changing continuous cross-sections, made by a single transit of a single cube, constitutes for the latter—the *individual*—one physical incarnation, and for the figure itself—the *personality*—its single and sole existence, we are now in a position to consider life from the point of view of such a personal consciousness confined within a changing perimeter—its life vehicle—and limited to the two dimensions of a plane—its world.

PLANEWISDOM

Suppose we seek out the most intelligent inhabitant of this "Flatland," and limiting our perception in the way his would be limited, share his consciousness. This Flatlander we will name Planewisdom, since he is of both an observant and a reflective turn of mind. Looking upon life and the world, the perpetual flux of things would perhaps first and most impress him—no rest, no stability, no finality anywhere. He would perceive himself to have come mysteriously "out of the nowhere into the here," quickly developing lines and angles, doomed presently to wane and disappear; and all of this kind would seem to be afflicted with a similar destiny—birth, a short perturbed existence, ending in dissolution. Of course Planewisdom would not have accurate knowledge of the true nature of all his fellow plane-beings, for this knowledge could come only from the standpoint of the higher-dimensional space; but in his transit he could examine the enclosing lines and angles of his fellow plane-beings and from these form some approximate notion of the distinctive characteristics of each. He would conclude that although every man was different from every other, they all had characteristics in common. All passed through certain recognizable phases in approximately the same space of time: a period of increment or growth, the attainment of a maximum, a phase of diminishment ending in disappearance. He would observe that some developed more varied outlines than others, making them lead troubled, irregular lives; such as began as a point being more "unhappy" than those which began as a line, because in the latter case, though one dimension changed, one remained constant. In those rare instances in which all four boundaries appeared at once and of equal length, a serene and unperturbed existence was the result.

Now, suppose that Planewisdom, wishful for this untroubled, equable life, should attempt to develop his changing irregular polygon evenly and symmetrically like the square. Failing in every effort to modify his perimeter, he might conceive the idea that a change of contour could be brought about only by a change of consciousness. He would recognize a distinction between his body and his consciousness. Though his body was confined to the plane world by the conditions of its existence, this limitation need not necessarily extend to his consciousness. This consciousness he would realize to be within him, yet who could say that it was on that account confined to the plane which constituted his world? What if there were an unknown direction, at right angles to the two known to him, in which his consciousness were capable of rising—the third dimension, in point of fact? Imbued with this idea, he might succeed, by an act of faith and by an effort of will, in uniting, after a fashion, his personal consciousness with his cube consciousness, and endowed with a new and mysterious

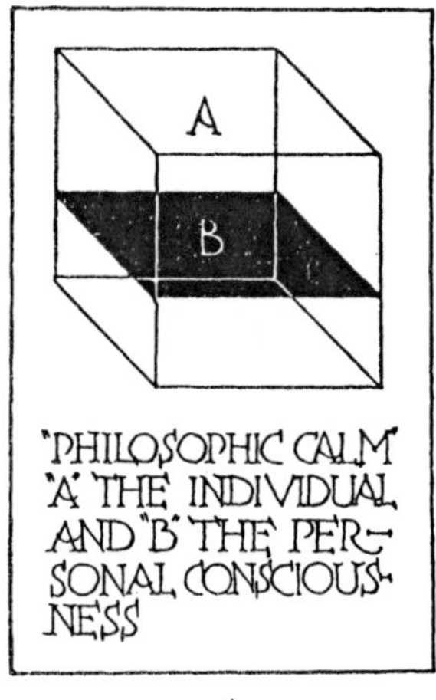

"PHILOSOPHIC CALM" "A" THE INDIVIDUAL AND "B" THE PERSONAL CONSCIOUSNESS

4

power, he might react upon his perimeter in such a manner as to change it from an irregular polygon into a square. This would produce a result in the higher world of which he could not but be unconscious; namely, that of bringing the vertical axis of the cube, his immortal body, perpendicular to the film. [Illustration 4].

PLANELOVE: PLANEBEAUTY

Planewisdom, being somewhat of a philosopher and a metaphysician, a "practical occultist" as well, his studies would tend rather to separate him from, than to unite him with, plane-humankind. Let us think of his neighbor, Planelove, as less intellectual, more emotional. Living so largely in his affections, the idea of withdrawing himself into his interior higher self, and so effecting a symmetrical development, would not even occur to him, and if it did would not attract. He could only achieve the symmetrical form in a different manner. Loving his fellow plane-beings, he would feel an *inclination* toward them, an attraction between his perimeter and theirs.

The fulfillment of this love would be in contact, juxtaposition. If one of his angles touched one of the sides of another, the resulting satisfaction would be ephemeral and slight, but the cleaving of line to line would effect a union not so readily nor so soon destroyed. Length of line being the gauge of felicity, the contact of the four sides of a square with the sides of four other squares would represent an absolute maximum. In the endeavor to attain the realization of this dimly felt ideal—that is, in multiplying his points of contact with his fellow plane-beings so as to prolong all of his lines equally and to the utmost—Planelove, in obedience to the universal law that one becomes that which he persistently admires and desires, would transform himself from an irregular polygon into a square. This change of outline would have the inevitable effect of altering the angle of inclination of *his* cube to the vertical, thus bringing about the same result achieved by Planewisdom through an intellectual process accompanied by a volitional effort; namely, a symmetrical development, a life free from mutations. [Illustration 5].

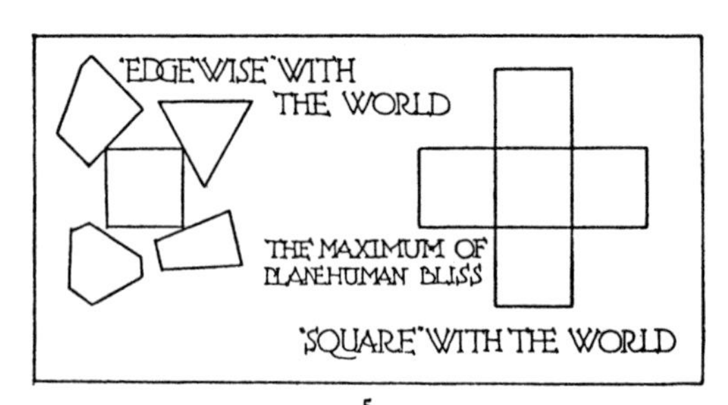

5

Let us imagine Planebeauty to have a personality still different from the other two. The thing which would trouble him in life would be, not so much its evanescence, its futility, as its dearth of ordered and formal beauty. The square would excite his admiration on account of its symmetry: four equal lines, four equal angles, and the persistence of this one most perfect form throughout an entire "life." By dwelling on the unique properties and perfection of the square, Planebeauty also would shift the center of his being in such a way that his higher, or cube self, would bring four of its sides vertical to the film world, and so his cube would trace out a square in the matter of that world.

THE THREE WISE PLANEMEN

Planelove, Planewisdom, and Planebeauty are "self-made" men of a plane world: they have become what they are by the way in which they have faced the world, incarnation after incarnation. At last, by learning to meet life *squarely,* they have arrived at a serene and equable maturity. Drawn together by the pursuits of a common ideal achieved by each in a different way, they *incline to,* and become *attached* to one another; for their lines being in contact, so also, without their knowing it, are the square faces of their cubes. This gives them a community of consciousness which others do not share; but they find themselves unsatisfied, for all their philosophic calm. What pleasure can a wise man take in a world of fools, they ask one another; a loving man in a world where people are indifferent to one another, an artist in a world so full of irregularity and imperfection. Even if all men were as they, still would the riddle of life be as far from solution as before? Their coming into the world would remain a mystery, and death would be the inevitable end of all. Is there no immortality, no rest, no peace, no knowledge, no perfection anywhere? Sadly they ask themselves these questions. Raising their plane-bound thoughts to the seeming void above them, they earnestly desire that some mighty teacher, perfect in wisdom, beauty, and compassion, may come and solve for them the riddle of their painful earth. [Illustration 6].

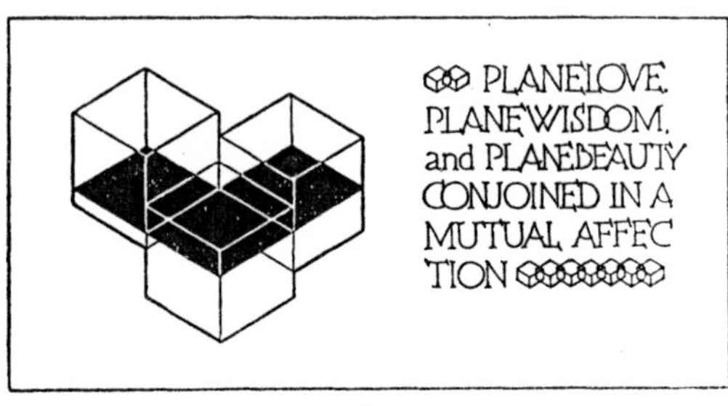

6

THE LIGHT OF THE WORLD

From its bright and shining sphere, a pure ray of the One Light gives heed to this prayer of the three plane-men. By reason of having passed through crystal prisms without number it has learned the rainbow's uttermost secret, and is therefore free from the necessity of refracting itself in substance, which, by impeding, reveals the latent glories of the one White Light. Looking down upon the cube, "like unto clear glass," it beholds, deep within, the insubstantial pageant of the film world in which men are but shadows of their higher, or cube selves, of which they are unaware, because their consciousness is centered in these lower-dimensional presentments of themselves. It takes note of the united desire in the hearts of three of these tiny and transitory figures. It is aware of their lonely struggle towards the only perfection that they know, and perceiving that the time has come for the Great Renunciation, it leaves the bosom of the

Father, the White Light, to show forth His glory in a world darkened by ignorance and death.

It enters one of the little crystal cubes, and shattering it along certain of its edges, folds its six faces down into the film-world so that they lie there in the form of a cross—a seven-fold figure, four squares lengthwise and three across. Thus it crucifies itself upon matter that by its broken body it may manifest to plane-men as much as they may apprehend of their higher, or divine selves. This body of Christos, though so poor and limited a vehicle for the divine consciousness within, is nevertheless glorious compared with the most perfect plane-human form, the square, for it is not one square, but six, each representing a different aspect of the Higher Self. [Illustration 7].

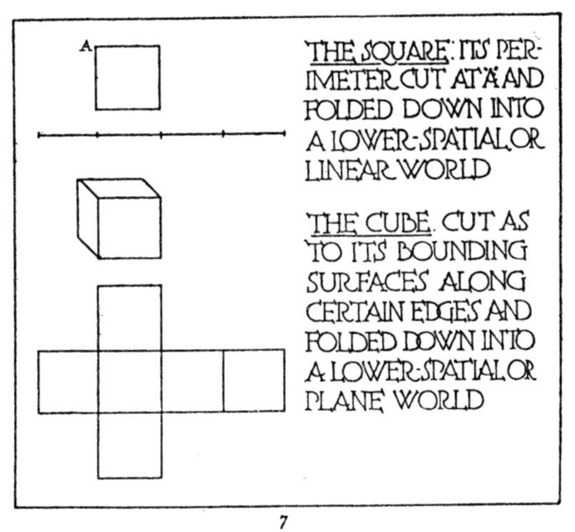

7

The three Wise Plane-men, having seen the descending ray as a star in the East, worshiped the incarnate savior of their world, at the place of his nativity.

Equipped with all knowledge, full of compassion, clothed in that transcendent body which had been "broken" for plane-humankind, this Christos of a lesser world in due course gathered disciples about him—square men all (save one)—to whom he taught the precious secret of release from birth and death. When his work was accomplished, he "ascended into Heaven," folded himself up in his higher-space form, and became once more a ray of the One Light, having committed the spreading of the truth to those whom he had instructed.

THE SERMON ON THE PLANE

One of his discourses has been preserved: rightly interpreted it will be seen to be in curious accord with the teachings of every world savior; for through all their utterances runs an eternal unanimity, the same in all ages, places, spaces. This is *The Sermon on the Plane,* preached to the plane-men by Him who was "crucified."

"Heaven is all about you: a city lying four-square, clear as glass and filled with light. Here your real, your immortal selves, have their true home. This world of yours which seems so substantial is but a mutable and many-colored film staining the bright radiance of this crystal heaven. Your lives are but tracings made by your immortal selves in this film world. How shall you learn the way to this heaven of light, the truth of this transcendent existence? I am the Way, the Truth and the Life. This is my body, broken for you. This cruciform figure formed by these six squares is not my immortal body; the squares are but the boundaries of it, folded down into a lower-dimensional world. When my mission is accomplished and I ascend again into heaven, I shall refold these squares into a single symmetrical figure, my heavenly body, a solid of the higher-dimensional space beyond your perception.

"Harken to the truth! Because these squares are solids of your world, it is hard for you to understand how they can be boundaries of a 'higher' solid. In order to understand it, imagine for a moment that your world, which is two-dimensional, is the higher space, or heaven world of a *one*-dimensional space. The perfected body of a plane-man is a square: suppose you wanted to give some idea of the square, which you are, to a consciousness limited to the one dimension of a line—a line-man. First of all, you would have to lose your solidity, forsake and forego your inner, or plane life, which for you is the only true life, and confining your consciousness to your perimeter, break it at one of its angles, and fold it down—straighten it out—into a one-dimensional space. Its four divisions, each one the boundary of one side of the square, would be, to the perception of the line-man, a solid of his space, and he would have the same difficulty in imagining them folded up into a single symmetrical figure that you have in imagining these six solids of your space to be the boundaries of a symmetrical solid of a space higher still. [Illustration 7].

"Each of you has this heavenly, or cube body, which you must think of as related to your physical or square body, as that is related to one of its bounding lines. The cube is the true individual, of which the square is but a single illusory and inadequate image. The individual expresses itself in countless of these personalities, each one a tracing of itself: the sum total of all possible tracings is the cube itself. Birth and death are illusions of the personality. For the cube they are not, since it did not begin its existence with its first contact with the film which is your world, nor will that existence cease when it passes beyond that world; neither does the changing cross-section which it traces, in thus passing, comprise or comprehend its life. Time and change are illusions of the personality. The cube knows neither increment nor diminution. All conscious cross-sections inhere within it—all possible forms of the film world. It is their revelation only which is successive, giving rise to the temporal illusion.

"Learn now the precious secret of immortality. The consciousness within the cube and within the square are one consciousness, and that consciousness is divine. It is possible, therefore, to identify your plane consciousness with your cube consciousness, and rise, by such means, into the higher-dimensional world. This is achieved by desire, by work, by knowledge, by devotion—but more than all by love, as you shall learn.

"Because each individual traces in the film world a different figure, determined by the angle at which it meets the film—by its *attitude toward life* —you are all under the illusion that each person is unique and singular, that some are better and some worse. But these differences are accidental: they do not exist in the heaven world, where all are God's children, and may become one with the Father. Live *uprightly,* love and cleave to one another. By so doing you will make vertical and parallel the axes of your higher, or cube bodies; and as the sides of your square bodies cleave together, so will the faces of your cube bodies coalesce. By loving your neighbor, therefore, you are "laying up treasure in heaven;" for two cubes can unite their faces only when the lines of their square sections are similarly joined. Love effects this junction.

"When cubes conjoin in mutual love the individual is transcended, the consciousnesses merge into one, and a larger unit is formed. This process may go on repeating itself, so that if love should become the universal law of life, the aimless drift of souls would cease, for all would enter the Great Peace at last. All having united into one great crystal cube, the Heavenly City, the film world would vanish. The White Light would shine unobstructed through the City of the Lord. [Illustration 8].

"It is thus that consciousness becomes self-conscious. It multiplies itself. Each unit, in its cube body, attains to a realization of its form and structure through the many tracings that it makes in physical matter (the film world), each transit being an incarnation, a personal life. The events of each life seem, to the personal consciousness, to slip away into nothingness, never to be recovered; but every experience of every film life, all of its contacts with other cubes, are indelibly impressed upon the higher body and by the cube-consciousness may be recovered at will, since all inhere in the bounding planes of the cube. For this reason, when cube consciousness is attained by the personality the memory of past lives is recovered. All lives may be lived over again as vividly as before: the indwelling consciousness has only to seek out in the boundaries of its cube body the particular point or line of contact with the film world in which the vanished event inheres. More than this, when any cube unites with any other, the indwelling consciousness of each, overpassing its normal limitation, is able to share in all of the past experiences of the other as though they were its own. By multiplying these contacts until all the cubes coalesce, each individual consciousness might share the experience of every other, from the dawn to the close of the cycle of manifestation. This is Nirvana, 'the Sabbath of the Lord.'

"These things I have spoken unto you that in me ye might have peace, that all may be one, and that they may be made perfect in one. My peace I give unto you."

THE INTERPRETATION OF THE SERMON

This elaborate paraphrase of familiar religious teachings needs no further elucidation if the reader has had the patience to follow it up, step by step, raising, as he did so, everything one "space;" that is, conceiving of man—the true individual—as a higher-dimensional entity manifesting itself in a three-dimensional space through and by means of the forms and conditions proper to that space.

To show how curiously in accord the higher space hypothesis is with the teachings of the Ancient Wisdom of the East, this essay will close, as it began, with a quotation from Madame Blavatsky's *The Secret Doctrine:*

"The real person or thing does not consist solely of what is seen at any particular moment, but is composed of the sum of all its various and changing conditions from its appearance in material form to its disappearance from earth. It is these 'sum totals' that exist from eternity in the Future, and pass by degrees through matter, to exist for eternity in the Past. No one would say that a bar of metal dropped into the sea came into existence as it left the air, and ceased to exist as it entered the water, and that the bar itself consisted only of that cross-section thereof which at any given moment coincided with the mathematical plane that separates, and, at the same time, joins, the atmosphere and the ocean. Even so persons and things, which, dropping out of the 'to be' into the 'has been,' out of the Future into the Past—present momentarily to our senses a cross-section, as it were, of their total selves, as they pass through Time and Space (as Matter) on their way from one eternity to another; and these two eternities constitute that Duration in which alone anything has true existence, were our senses but able to recognize it."

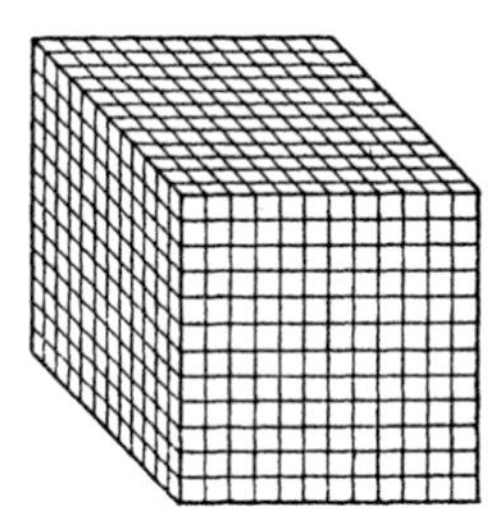

COSIMO is an innovative publisher of books and publications that inspire, inform and engage readers worldwide. Our titles are drawn from a range of subjects including health, business, philosophy, history, science and sacred texts. We specialize in using print-on-demand technology (POD), making it possible to publish books for both general and specialized audiences and to keep books in print indefinitely. With POD technology new titles can reach their audiences faster and more efficiently than with traditional publishing.

> ➢ **Permanent Availability:** Our books & publications never go out-of-print.
>
> ➢ **Global Availability:** Our books are always available online at popular retailers and can be ordered from your favorite local bookstore.

COSIMO CLASSICS brings to life unique, rare, out-of-print classics representing subjects as diverse as *Alternative Health, Business and Economics, Eastern Philosophy, Personal Growth, Mythology, Philosophy, Sacred Texts, Science, Spirituality* and much more!

COSIMO-on-DEMAND publishes your books, publications and reports. If you are an Author, part of an Organization, or a Benefactor with a publishing project and would like to bring books back into print, publish new books fast and effectively, would like your publications, books, training guides, and conference reports to be made available to your members and wider audiences around the world, we can assist you with your publishing needs.

Visit our website at www.cosimobooks.com to learn more about Cosimo, browse our catalog, take part in surveys or campaigns, and sign-up for our newsletter.

And if you wish please drop us a line at info@cosimobooks.com. We look forward to hearing from you.